Bill R

op

s

INFERNAL MACHINES

INFERNAL MACHINES

*The Story
of Confederate Submarine
and Mine Warfare*

MILTON F. PERRY

LOUISIANA STATE UNIVERSITY PRESS

BATON ROUGE AND LONDON

The paper in this book meets the guidelines for permanence and durability
of the Committee on Production Guidelines for Book Longevity of the Council on Library Resources. ∞

Louisiana Paperback Edition, 1985
94 93   4 5

Library of Congress Cataloging-in-Publication Data

Perry, Milton F.
  Infernal machines.

  Bibliography: p.
  Includes index.
  1. United States—History—Civil War, 1861–1865—Naval operations—Submarine. 2. Mines, Submarine—United States—History—
19th century. 3. Torpedoes—History—19th century. I. Title.
E596.P4 1985   973.7'57   85-9706
ISBN 0-8071-1285-2 (pbk.)

To Pop and Tanny

Despite the vast amount of literature concerning the American Civil War, there is no detailed account of the development and application of torpedoes and mines by the Confederacy. Some authors have included references to them as parts of larger works or in personal narratives, but, except for a few technical studies by military and naval engineers, the story has been untold. This is an attempt to rectify the oversight.

The subject of torpedo warfare was frowned upon as "cruel," "wanton," or "uncivilized" by many contemporaries of the Civil War. This attitude, plus the lack of extensive documentation, is probably the reason that the telling of the story has so long been avoided. Yet the weapons gave rise to a whole new dimension in military technology. The intricacies of the mines themselves, the methods of delivering them, and the counterweapons they engendered far exceeded any of the plans of the inventors or weaponers. True, if the Confederacy had not adopted the use of mines, some other nation would have, but this does not diminish their historical impact. There still remain the exciting displays of courage found in the adventures of Matthew Fontaine Maury, Francis D. Lee, Hunter Davidson, James A. Tomb, Beverly Kennon, and their comrades.

I was born near the confluence of the Roanoke and Chowan rivers with Albemarle Sound, and grew up on the shores of Hampton Roads. My ancestors probably heard the cannonading of the naval battles and mine explosions in those areas. I swam in the muddy Roanoke near a "Civil War gunboat wreck." Curiosity prompted me to study the war there. Later, as a member of the staff of the West Point Museum, in which is preserved the only large collection of Confederate torpedoes, I found there was no full history of the weapons.

The assistance of many persons made this project possible. I want to acknowledge especially the many notes of encouragement and the review of the manuscript by Mr. Philip Van Doren Stern, distinguished author of a number of excellent Civil War studies. Guy E. Bishop, Jr., a friend of many years, and William L. Philyaw (who drew the maps of the Charleston area and Mobile Bay) perused the manuscript and offered valuable suggestions. Richard Lester of London willingly ferreted out references in British sources. Stan Olsen, Jim Hillard, Gerry Stowe, and William Luckett, gave me pointers and suggestions.

Peggy Smith, Cathy Mullen, Marcy Warren, and my wife Barbara labored valiantly with notes in my nearly illegible hand or with my "peck" typing. Theirs, perhaps, was the most difficult task of all.

M. F. P.

CONTENTS

ILLUSTRATIONS

INFERNAL MACHINES

CHAPTER I

MAURY'S TORPEDOES

On Sunday, July 7, 1861, the gunboat U.S.S. *Pawnee* was patrolling the choppy waters of the Potomac downriver from Washington. Aloft, and on deck, her crew was searching for small boats crossing the river or for activity in the Confederate batteries protecting the mouth of Acquia Creek, the route to the railroad for Richmond. A sharp-eyed lookout drew attention to a pair of black specks bobbing with the tide some two hundred yards from the ship. Commander Stephen C. Rowan ordered his vessel stopped and sent her consort, the *Resolute,* to investigate. A cutter, directed by Master William Budd, was launched and took the as yet unidentified objects in tow. Curious sailors peered down as Budd tied onto the ship. They saw a pair of large barrels connected by a long rope. Suspended beneath the barrels were two iron containers filled with powder and fitted with waterproof fuses.

Rowan recognized these contrivances as a pair of the "explosive machines" the Confederates were rumored to be perfecting. Budd

3

gingerly pulled them toward the shore after pouring water into the air holes to extinguish the fuses. On the way one sank, but the other was rescued and sent to Commander John A. Dahlgren of the Union Navy's ordnance bureau.[1]

This was the first appearance of the "torpedo" in the Civil War. Before hostilities ceased three years and nine months later, this weapon was to destroy or damage twoscore Federal ships and kill hundreds of men. Torpedoes (they are called "mines" today) were to send more Union vessels to the bottom than were all the warships of the Confederate Navy—in the James, at Charleston Harbor, in the Red River in the West, and during Farragut's famous encounter at Mobile Bay.

Although millions of dollars were poured into the construction of ironclads, ship-to-ship encounters were rare, and not one of these vessels succeeded in sinking an opposite number. Torpedo service grew so rapidly that by 1863 it became prudent in many areas to keep one's monitors and rams anchored, ready for action, but surrounded by nets and rows of obstructions and circled by small boats that examined the water, using calcium lights at night. These elaborate precautions were devised as the most effective means of protection against floating mines or attacks by swift, low torpedo boats which would be developed later in the war. Despite such measures, some fifty ships were eventually sunk or damaged by mines. Forty-three of these were Union, a figure that embraces the destruction of four monitors. Only a single Confederate vessel was sunk by Federal torpedoes—the ram *Albemarle*—a naval exploit that became one of the best-known of the war.

Torpedoes were not entirely new to warfare, especially in America. During the Revolution, David Bushnell had devised a keg torpedo which was made famous by Francis Hopkinson's satirical poem "Battle of the Kegs." Robert Fulton and Samuel Colt both became interested in such explosives but turned toward other pursuits when their experiments were not well received, though both succeeded in sinking target ships. Floating mines were used at Canton, China, in 1857–58, and by the Russians during the Crimean War. Desultory experiments had been conducted elsewhere, but few except the inventors seemed to take them seriously.

Among those who did was America's distinguished scientist-sailor-oceanographer Matthew Fontaine Maury, who, for his charting of the ocean currents, has become known as "The Pathfinder of the Seas." The fifty-four-year-old Virginian resigned his commission in the U.S. Navy on April 20, 1861, after thirty-five years of service, and cast his lot with the Old Dominion because, as he said, he had been raised "in the school of states-rights." [2]

The stocky, balding scientist had reached the decision to terminate his long and honor-filled career only after much introspection. Lame from an accident in 1839, he had served in the Navy's hydrographical office and had published his findings in several books, among them *Physical Geography of the Sea,* which described the nature of the Gulf Stream. Impulsiveness was alien to his character; his choice was a deliberate one. Virginia, he wrote, "gave me birth; within her borders, among many kind friends, the nearest of kin, [and] neighbors, my children are planting their vine. . . . In her bosom are the graves of my fathers. . . ." This, to him, was "the path of duty and honor." With his large family he left Washington and moved to the house of a relative in Fredericksburg. [3]

He was called to Richmond and on April 23, 1861, was placed on the governor's advisory council to make recommendations for the protection of the state's waterways. It was at this juncture that he turned to the possibility of utilizing torpedoes in the defense of the South. The Southern states had but few ships; even if they had had more, the three thousand miles of coast, indented with numerous bays and rivers, posed a difficult defensive problem. Torpedoes, he reasoned, could take the place of nonexistent warships and present a relatively inexpensive ring of defense.

Maury seems to have come to these conclusions about the time of his arrival in Richmond and tried to convince the authorities of his views. However, those to whom he talked opined that this was an "ineffectual and unlawful warfare." At the home of his cousin Robert H. Maury on fashionable Clay Street, he experimented in his bedroom with tiny cans of powder and a large washtub filled with water, splashing the furnishings and frightening the family with sharp explosions. Satisfied that his plans were the basis for a practicable undertaking, Maury had a large oak cask built to his specifications by Talbott & Son on Cary Street. Two barrels of

5

gunpowder were placed at his disposal by Governor John Letcher, and a full-scale test was planned on the James River off Rocketts Wharf.

It was a hot, sultry day in mid-June. A host of dignitaries—including the governor, the Confederate Secretary of the Navy, and the Congressional Committee on Naval Affairs—and Maury's wife and children crowded the wharf and bank. Taking his son, Maury rowed out into midstream from the C.S.S. *Patrick Henry*, floated his torpedo, set the percussion trigger, and carefully unrolled a rope lanyard. At a safe distance, his son yanked the cord. "Up went a column of water fifteen or twenty feet," wrote his son Richard. "Many stunned or dead fish floated around; the officials on the wharf applauded and were convinced. . . ." Actually, Maury had hoped to demonstrate something even more spectacular—an electric torpedo fired by a spark passing through a long, insulated cable—but the wire was not to be had.

The "doubting Thomases" were silenced, and Maury was promoted to "chief of the Sea-Coast, Harbor and River Defenses of the South," with the rank of captain, "abundant" funds ($50,000, which seemed abundant then), and a group of eager assistants under his command.[4]

He lost no time in planning the destruction of the nearest Union warships. Five vessels lay at anchor in Hampton Roads, presenting a tempting target; these included two flagships, the *Minnesota* and the *Roanoke,* commanded by old acquaintances. Maury gathered a small crew and went secretly by train through Petersburg, along the southern shore of the James to Suffolk and Portsmouth, alighting at Norfolk on Friday, July 5, 1861. It was only a few miles to Sewall's Point, overlooking the Roads, the location of Confederate batteries which guarded the mouth of the Elizabeth River.

That night and the next, they attempted reconnaissance in a small boat, but a picket steamer "flying round and round" kept them away from the anchorage. On Sunday, while watching the Federals through his spyglass, Maury, an intensely religious man, saw the church flag raised above the Union colors and suffered twinges of conscience and a wave of nostalgia for the many times he too had been under these same two banners. Nevertheless, he set out at ten o'clock that night, leading five skiffs, each carrying a torpedo with thirty fathoms of lanyard. The intention was to join two casks with line and let the tide carry them across the enemy's bows. If everything went as planned, the current would sweep the torpedoes along the sides of the ships, and the resulting strain would trigger the fuses.

"The night was still, clear, calm and lovely," said he of the experience. "Thatcher's Comet was flaming in the sky. We steered by it, pulling along in the plane of its splendid trail. All the noise and turmoil of the enemy's camp and fleet was hushed. They had no guard boats of any sort out, and as with muffled oars we began to near them, we heard 'seven bells' strike." [5]

To reach the fleet it was necessary to row about three miles. Carefully checking their position and the flow of the water, they passed the "Middle Ground" near the center of the Roads and laid a course toward Newport News on the western shore, taking care to move ahead of the fleet toward the mouth of the James. Maury probably timed his movement so that the tides ran with him.

His daughter Betty recounted the rest of the story in her diary:

After putting the magazines under one ship, the boats . . . were ordered back, and Papa went with the other two to plant the magazines under the other vessel.

They then rowed to some distance and waited for the explosion, but it never came. Thank God, for if it had Pa would have been hung long before now.

Pa thinks he can account for the failure, and could rectify it very easily.

Says he was very much struck with the culpable negligence of the enemy. That he could have gone up and put his hand on those vessels with impunity.

There was a little mechanic who went . . . from Richmond with express stipulation that he should not go in the boats with the expedition. But at the last minute he got so excited and interested that he begged and implored to be allowed to go.

It would have been a grand success.

When Maury later recounted this episode to his family, he was asked in mock seriousness if he would have rescued Lieutenant Austin Pendergast of the *Minnesota*. Pendergast had been a member of the board that "turned" Maury from the U.S. Navy, even though he had formally resigned six days before the board met. The captain thought a moment before making his embittered reply: "I should like to have seen him and asked him if he thought I was fit for active service now!" [6]

In her fears for her father's safety, Betty reflected her mother's disapproval, and so did another daughter, Mary: "My mother was not in favor of the use of the submarine torpedoes and thought it barbarous to blow up men without giving them a chance to defend themselves." Maury calmly replied that by these measures he hoped to shorten the war. [7]

The Federals were completely unaware of the attack. Until one of the torpedoes was seen adrift in Hampton Roads two weeks later, they had not the slightest inkling of it. Maury's participation did not become known until after the war. When the captured mine was examined, it was found that the powder was wet from a leak, a possible explanation of its failure to explode. [8]

This effort seems to have been part of a well-timed, double-barreled attack with torpedoes, for while Maury was busy in Hampton Roads, Lieutenant Beverly Kennon, Jr., C.S.N., was trying to sink the *Pawnee* in the Potomac. Since the publication of Virginia's ordinance of secession, this well-manned ship—one of the few sound warships in the United States Navy—had hovered off Acquia Creek, occasionally dueling with artillery which the Confederates had sunk in pits along the shore. Whether or not Kennon was dispatched by orders from Maury remains a mystery. Betty Maury wrote that her father was not aware of the other attack, but the timing raises the question. In addition, Captain Maury directed precisely the same kind of simultaneous attacks later that year. In any event, neither of these particular endeavors succeeded. [9]

The other attacks were planned in October, 1861, although Maury's family begged him to withdraw from these hazardous undertakings. However, the Navy Department intervened by placing him on "special duty which will interfere with his orders relating to the submarine battery. . . ." In his place went Lieutenant Robert D. Minor, who had been sent to Maury by Confederate Secretary of the Navy Stephen C. Mallory. [10]

At 8 A.M. on October 9, Minor boarded the *Patrick Henry* off Mulberry Point in the James. She was a former New York and Old Dominion Line steamer that had been used in the June torpedo demonstration and was destined to become the school ship of the Confederate Navy. Minor met her skipper, Commander John R. Tucker, and his executive officer, acquainted them with the plan, and secured two volunteers to help him—Master Thomas L. Dornin and Midshipman Alexander M. Mason. He spent the afternoon arming the gray-colored weapons and explaining their operation to his assistants. At this point Minor discovered that he had only 392 of the 400 pounds of gunpowder he had intended to use. This meant that the explosion of each torpedo would be somewhat weaker than he had calculated.

Commander Tucker got under way at sunset with shrouded lights, moving slowly downstream on the outgoing tide. The *Patrick Henry* rolled heavily in the swells kicked up by a nor'easter—weather ideally suited for the mission, said Minor.

Somewhat later, dark shapes of the enemy ships were seen through the rain and mist off Newport News. The Confederate vessel anchored quietly, using a hawser instead of a chain. Minor wrote:

The boats were then lowered, the magazines carefully slung, buoys bent on at intervals of 7 feet, and, when all was ready, the crews, armed with cutlasses, took their places and were cautioned in a few words by me to keep silent and obey implicitly the orders of the officers. Acting Master Dornin, with . . . Mason, took the left side of the channel, while I took the right with Mr. Edward Moore, boatswain . . . to pilot me. Pulling down the river some 600 or 700 yards, the boats were then allowed to drift with the rapid ebb tide, while the end of the cork line was passed over to Mr. Dornin and the line tautened by the boats pulling in opposite directions. The buoys were then thrown overboard, the guard lines on the triggers cut, the levers fitted and pinned, the trip line made fast to the bight at the end of the lever, the safety screws removed, the magazines carefully lowered in the water, where they were well supported by the buoys, the slack line . . . thrown overboard, and all set fairly adrift within 800 yards of the ship and 400 yards of the battery on the bluff above the point. So near were we that voices were heard on the shore, and Mr. Moore reported a boat about 100 yards off. . . . Pulling back a short distance and hearing no explosion we returned to the ship, which we found cleared for action and ready to cover us in the event of being attacked. . . .[11]

No sooner had the boats been hoisted than the lookouts aboard the *Patrick Henry* observed signal lights at Newport News flashing furiously and the Confederates thought they had been discovered. The anchor was raised and the *Henry* made upriver at top speed, reaching Mulberry Point at 12:30 A.M.

Daybreak found Minor in the crosstrees looking toward the Federal anchorage where he could make out a pair of ships in the same position as on the previous evening. Later, he and Tucker rowed down to the mouth of the Warwick River where they could see better. Both of the enemy appeared unharmed. The torpedoes had failed to ignite. The United States squadron had detected the Confederates, but had drawn the wrong conclusions. They believed they had been attacked by a Confederate submarine, which they confidently thought was later sunk.[12]

That summer Maury discussed mines with Lieutenant Isaac N. Brown, C.S.N., a sandy-bearded veteran of more than twenty years with the U.S. Navy, and an old comrade. Brown had been assigned to work with Major General Leonidas Polk in building defenses along the Mississippi River. Brown described his problem and Maury proposed setting out electric torpedoes in the river. After studying maps and examining various sites, they selected Randolph Bluffs just above Columbus, Kentucky, as the best spot "from Cairo to New Orleans" for the new system.[13]

Before the torpedoes could be designed, more detailed information was needed about the enemy's ships. Rumors had seeped down the river of various new types of vessels the Federals were building at Cairo and Cincinnati. On July 20 Secretary of the Navy Mallory instructed Brown to send someone who knew ships to Cincinnati in order to learn the "character of the vessels."[14] Polk requested that Secretary of War Leroy P. Walker have Captain Maury supervise the positioning of the submarine batteries in person, but apparently he was not able to go. Instead, he seems to have planned to make the components in Richmond and send them to Columbus with directions for assembling. He wrote Brown: "Assuming that you would want them for the river at Columbus, I have been pushing them forward. I shall have, by the end of the week, six all ready to be filled and planted. An unexpected run of good luck has enabled me to do this. I have twenty-five underway, all of which, if, in spite of the driftwood and other habits of the Mississippi, they can be made to answer . . . are at your service."[15]

A Mr. A. L. Saunders of Memphis wrote that he, too, wanted to help and sent a number of torpedoes of his invention. Beverly Kennon, who had been building torpedoes and other ordnance materiel at New Orleans, took some to Brown and helped him put them in the river.[16] This was the first combat electric mine station to be established in America and, probably, in the world. Maury's cylindrical containers were buoyed across the river. They were three feet long, were pointed at one end, and weighed about fifty pounds. Precious lengths of insulated wire connected them with galvanic batteries and telegraph key contacts hidden in caves along the thickly wooded bluffs.

Others were planted on land. These were squat iron castings with handles, resembling covered pots. Their lids were secured by eight bolts, and beneath were smaller wooden boxes that protected the holes through which the wires from the batteries passed. Inside were 4-pound artillery shells filled with canister and grapeshot, and two bushels of gunpowder. These mines were planted along the two roads leading into Columbus from the north. Six wires ran from clusters of mines beside the Clinton Road and a hill near it to one cave, and a similar arrangement guarded the Elliott's Mill Road atop the river bluffs. It is not certain which inventor had devised this scheme.

Rumors of these dangers reached the Federals while they were preparing to move on Columbus and on Belmont, across the river in Missouri. In March, 1862, they carefully investigated the river and captured the system without exploding a single mine.

The land mines were harder to find. When the Confederates abandoned the river defenses, the telegraphic switches were deserted in working condition, and no attempt was made to detonate the mines beneath the feet of Union soldiers. They were discovered by Captain W. A. Schmitt and his company of the Twenty-seventh Illinois Infantry. Schmitt's men spotted a series of new ridges in the earth, indicating that shallow trenches had recently been filled. Tracing them, they came upon the paraphernalia in the caves. At the ends of the ridges they uncovered mounds which revealed the explosives.[17]

One observer claimed to have understood that there were from seventy-five to one hundred mines of various types at Columbus, and another put the count nearer to four hundred. There were equally conflicting stories as to the kinds of torpedoes—one reporter listing at least four distinct types. In addition to the electrical varieties, he described a cylindrical percussion water mine and another "with some brass arrangements which it is difficult to designate." Whatever they were, they did not work. The powder was wet.[18]

CHAPTER II

THE SUBMARINE BATTERIES

Though he had directed two attacks with mechanical mines, Maury was primarily interested in electric torpedoes and, after sending his first products to Columbus, concentrated wholly upon improving this type, leaving the future development of mechanical mines to his assistants. Obviously, insulated cable would be in constant demand, but there was none in quantity in the Confederacy. He therefore dispatched an agent to the nearest source, New York City, to buy telegraph wire from the enemy. This attempt failed; just why, no one knows for certain. Southern records mark it simply as having been "foiled." It is worth noting that at about the same time La Fayette C. Baker, head of the Union Secret Service, helped trap a Southern courier in Philadelphia who was to have gone to New York "to get some telegraphic material." [1]

A call was made on that bulwark of the South, its women, for donation of India rubber coats and overshoes. Many were sent, but they were not enough to wrap more than a few hundred feet of cable. Actually, many thousands would be needed for extensive mine laying.

Electrical science was then in its infancy. Maury turned to the South's leading expert, Dr. (Major) William Norris, chief signal officer. Though he originally "depreciated" the scheme, Norris eventually provided much technical advice as well as a supply of wire which had been a Federal underwater telegraph cable laid by the U.S.S. *Hoboken* from Fort Monroe to Eastville on Virginia's Eastern Shore. The wire had been recovered by alert Rebels after being thrown upon the beaches of Willoughby Spit near Norfolk by a storm on February 25, 1862. Though it was frayed and broken from its buffeting on the floor of the bay, Maury considered it "a God-send . . . just what we wanted."[2]

Several influential persons inspired, no doubt, by Maury's distinguished reputation and persuasive manner, came to recognize the importance of the mines. Navy Secretary Mallory and President Jefferson Davis became his enthusiastic backers. Davis was presented with a resolution by the House of Representatives "in regard to the protection of our principal cities from iron-plated vessels by . . . obstructions and submarine batteries. . . ." Anxious because of the appearance of the little U.S.S. *Monitor,* the House asked if further monies were required to spur the work and Davis' prompt reply reflects a long discussion with Maury:

Submarine batteries have been and are being prepared, and . . . no additional appropriations for these objects are considered to be needed. Until recently the character of the enemy's iron-plated vessels was not well enough known to arrange obstructions specially for them, but the same principle obtains [as with wooden ships] and the obstructions already prepared can be strengthened when necessary. For the want of insulated wire we are deprived of that class of submarine batteries exploded at will by electricity, which promises the best results. Experiments upon several kinds of such as are exploded by impact have been in progress since an early period of the war. These torpedoes can be rendered harmless by the enemy in most cases by setting adrift floating bodies to explode them, as is said to have been done on the Mississippi River, and as they cannot be put in place so long as all the channels are required for use by our own boats no great degree of importance is attached to them. They may serve, however, to gain time by making the enemy more cautious; and most of our seacoast defences have already received, or will as soon as practicable receive, a certain supply of them.[3]

Precious little equipment to make the electric mines was at hand. For waterproof tanks or, as Maury preferred to call them, "magazines," the experimenters rounded up several old iron locomotive and steamboat boilers, cleaned them of rust, patched the holes, and filled them with gunpowder. A Wollaston battery was the only one available. (Later they borrowed a Cruikshank battery from the University of Virginia.) The Wollaston consisted of banks of cells in which were eighteen pairs of 10- by 12-inch zinc plates immersed in thirty-six gallons of sulphuric acid; it could generate enough current to explode a single charge only a few hundred feet away. "One can imagine the clumsiness and inefficiency of such means," reflected Maury's chief aide, Lieutenant Hunter Davidson, some years later.[4]

A second system of torpedoes was better accommodated to this primitive equipment. This one was similar to that used at Columbus and employed iron magazines of 70 to 160 pounds each, anchored in clusters or, as Maury designated them, "ranges."[5]

All spring they worked and experimented with these weapons. When the Federals began their Peninsular Campaign coupled with a naval expedition up the James, Maury wrote Secretary of War George W. Randolph on May 1, 1862: "I beg to call your attention to our river defenses, and to say that the most effectual way of keeping off the enemy with his shot-proof vessels is to mine the channelways, and blow up by means of electricity when he attempts the passage."[6]

The propeller tug *Teaser* was assigned to the "Submarine Battery Service" and with the assistance of a kinsman, Lieutenant William L. Maury, Lieutenant Davidson, and several others, Maury began to establish the defenses. One used four converted boilers placed in two ranges. Each of the mines held some 1,500 pounds of powder and were put down near the batteries on Chaffin's Bluff below Richmond. The other, consisting of fifteen smaller torpedoes, was placed below the bluff.

They are arranged in rows, those of each row being 30 feet apart. Each tank is contained in a water-tight wooden cask, capable of floating it, but anchored and held below the surface from 3 to 8 feet, according to the state of the tide. The anchors to each are an 18-inch shell and a piece of kentledge. . . . The wire for the return current from the battery is passed from shell to shell along the connecting rope, which lies at the bottom. The wire that passes from cask to cask is stopped slack at the buoy rope from the shell up to the cask. . . . The return wire is stopped in like manner down along the span to the next shell. . . .[7]

Ashore were twenty-one Wollaston batteries in sheds, hooked up so that several torpedoes in one range could be exploded at once. Before they could be tested, a spring freshet roared down the river and washed the larger tanks away. In spite of intensive grappling, they were never located.

In their "factory" Lieutenant Maury and his assistants went about "proving" the tanks and putting them in the gray casks. Spares were kept at hand so that the smaller mines could be readily replaced if washed away.

The lack of adequate facilities in the Confederacy imposed limitations upon serious scientific research. This drawback, together with Captain Maury's penchant for risking himself in attacks upon the enemy, was no doubt partially responsible for the Navy's decision to send him to England. Many other duties were placed upon him there, including the purchasing of warships. He could also conduct his studies in well-appointed laboratories with the assistance of trained personnel. In June, 1862, the orders were published. The result was that he would spend the rest of the war in Europe, where he developed a highly sophisticated system of electrical mines.[8]

Lieutenant Hunter Davidson was appointed his successor. His duties were "devising, placing and superintending submarine batteries in the James River" and exercising "discretion as to the ways and means of placing obstacles of this and of any other character to oppose the enemy's passage of the river."[9]

Why was this young man selected to replace a world-famous scientist who had just begun a most promising series of military experiments? Davidson, a graduate of the United States Naval Academy's second class, had served for nearly twenty years before joining the Confederacy in April, 1861. (He is carried on the Navy's rolls as being "dismissed" on April 23 even though, like other Southerners, he had submitted a formal resignation.) His qualifications were well known to Maury, who gave him a fine recommendation for having helped him "with a most hearty good will." President Davis later characterized Davidson as "an officer vigilant, fruitful in expedients, quick to perceive the defects incident to the application of a new engine in war, willing to admit errors, prompt in devising remedies, full of enterprise and intrepid in execution." It is probable that he was put forward by Maury, but the correspondence has not been found. Davis admitted that "the choice was fortunate." [10]

Davidson began by making a thorough study of the Maury torpedo system. "The fact that there was no *practical* result from his experiments," he wrote of Maury, "is due simply to the want of time to organize his forces and collect material, though his experiments served to mark some of the shoals on the way, if not the channel to success." [11]

Davidson suggested an almost completely new system, and received Secretary Mallory's approbation upon the presentation of his plans. But not all of those with whom he came into contact were so encouraging. Acquaintances begged him to give it up, jested, and at times even sneered at the project. They seemed to think he could do better by serving on a proven weapon—for example, an ironclad. When he called on Colonel Josiah Gorgas, the army chief of ordnance, for fine-grained rifle powder, the soldier would not even receive the torpedo officer. The impasse was resolved only by a direct order from the Secretary of the Navy. So low were mines held in the minds of various individuals that they refused to associate with anyone who worked with the devices.

Iron containers of three-quarter-inch boiler iron, especially designed for the purpose, were prepared by the Tredegar Iron Works in Richmond, while Davidson and his crew experimented with methods of ignition. Again the lack of batteries was a great drawback. Those used by Maury had been admittedly unsuited to the work, but they were the only ones available at the time. Davidson managed to get a Boynton galvanic battery and some submarine cable from England, and his electrician, R. O. Crowley, modified it, developing a battery with components so small that it is said a single man could carry them. Acids were obtained by requisitioning the meager supplies retained by druggists.

A pair of cables led to each of the powder-filled tanks. Inside, at the center, they were connected by a fine platinum wire that passed through a goose quill of fulminate. When the contacts were joined ashore, heat generated by the platinum exploded the fulminate and set off the mine in a fraction of a second. A test upon a ten-ton sloop was completely successful, "destroying the boat and throwing a column of water 30 feet in diameter about 45 feet high," all with only seventy-five pounds of explosives. A 200-pound mine raised a 60-foot column and tossed fragments more than a hundred feet into the air.

This work occupied Davidson and his men the whole summer of 1862. By late October they were about finished, and Davidson told a friend:

I am getting on slowly here with the submarines. I shall soon have about 12,000 pounds powder down at different stations on the river.

My later experiments prove that powerful galvanic batteries can be relied on to act with unerring certainty at the distance of a half mile under water. This is the way we should obstruct all our rivers if sufficient powder can be got. I don't believe you can find a Yankee to risk a blowing up.[12]

One of the several delays was occasioned by the enemy. On July 4 the Union gunboat *Maratanza*, which was on a reconnaissance up the James with the *Monitor*, surprised and captured the little *Teaser* after a short, sharp engagement. On board the tug was its mine-laying equipment including wire, batteries, powder, acid, and, in addition, a multicolored deflated balloon "made up of old silk frocks." The lighter-than-air craft was to be used in spying out Union positions at Harrison's Landing. The *Teaser* was promptly replaced with another tug, aptly christened the *Torpedo*.[13]

After establishing several stations on the James above Dutch Gap, Davidson led a distinguished group, including President Davis, General Robert E. Lee, and Secretary Mallory, on an inspection of the installations. So confident were they of the explosives, said Davidson, that large numbers of troops were immediately withdrawn for operations in other areas, "it being well understood that the Union armies could not advance without the assistance of the Federal squadron." [14]

Another station was established along the Rappahannock River not far from Port Royal, Virginia, using three torpedoes. It was a failure. Davidson's aides believed that no Federal ships ventured near the torpedo fields because a slave had disclosed the locations—an excuse that was to be used countless times, and was sometimes true. When the Army of the Potomac prepared its assault upon Fredericksburg in December, Davidson decided to remove the station lest the enemy occupy the peninsula between the York and Rappahannock rivers. He sent electrician Crowley to retrieve all the material he could. It was, wrote Crowley, "a hazardous undertaking" because the station lay in a no-man's-land, and a Union outpost was known to be just across the river. Crowley reached it at sunset and closed the gap. A "tremendous explosion" resulted, erupting in great columns of water tinted a delicate rose by the waning light. Collecting wires and batteries, the Rebels quietly made their way back to friendly territory.[15]

CHAPTER III

RAINS'S SUBTERRA SHELLS

It was quiet, too quiet, in front of the lines at Yorktown. Perhaps that camp rumor was right after all. Maybe the Johnnies *had* skedaddled and left Yorktown for the taking. These were the hopes of the men forming detachments of cavalry sent out to probe the roads leading to that historic old Tidewater Virginia town. Spring there comes early and beautifully, but the soldiers paid little heed to the blossoms of fruit trees or the birds singing in the branches that morning of May 4, 1862. Not a single shot had been fired, not a person seen. Suddenly the ground beneath one of the horses erupted with a violent explosion, lifting animal and rider, throwing them some distance to one side.

This was not artillery. No one had heard the whizzing of a projectile or the bark of a cannon. The men spread out and examined the ground. Their findings verified the report of a deserter who had entered the lines earlier that morning: shells with triggered fuses were buried among and in front of the fortifications.[1] Another column heard a "stunning report" ahead. A large hole, smoking and rank with acrid gunpowder fumes, came into view. At the roots of a tree lay a wounded soldier. Two more nursed lesser injuries while staring at the mangled body of a fourth.[2]

The word spread quickly. As the army entered the town on May 5, soldiers thought they saw torpedoes everywhere. Their letters and diaries reveal a newfound fear—fear of a weapon they could not see or hear, a weapon that lay dormant and concealed, causing death at the slightest touch. This was different from battle; in battle a man knew what his enemy was, could hear him and often see him. This was the unknown.

One soldier reported: "You could not tip over a barrel or anything else, but what had a string attached to a big shell or . . . torpedo, that would kill five or six men every time they did anything or moved anything. Wherever you could see dirt thrown up loosely, look out for your feet, or else they would be catching in some string an inch under the dirt, and then shells would explode."

"The gate to the fort stood open," reported another. "A heavy shell was planted there which the opening of the gate would have exploded. Our Army declined to enter. . . ."

A comrade put in his journal that "in some places you will see an overcoat laying on the ground, but it will not do to pick it up, for it is attached to a string leading to a fuse containing powder, so when . . . [it] is picked up it . . . explode[s]. . . . Torpedoes," he continued, "are covered with dirt in the street . . . in fact the whole place is mined, and . . . small flags are placed near the infernal articles to give . . . warning." [3]

An officer stated that "thickly studded all along the road . . . are . . . these fatal iron fuses, whose touch is death; but every inch of the road is carefully scanned. . . ." [4] Charles A. Phillips of the Fifth Massachusetts Battery of Artillery wrote on May 6:

A blood stain on the ground where a man was blown up . . . and a little red flag ten feet from it, admonished us to be careful. The rebels have shown great ingenuity . . . for our especial benefit. One house seems to have been the particular object [of] shells placed in all convenient spots. Under a table in the corner of the room was . . . a coffee pot . . . tied by a small thread to a weight hung directly over the cap

of a 10-inch shell, so that the weight would fall as soon as the . . .
pot was moved. Then the cellar floor was paved with similar machines at the foot of the stairs. . . . There is one room which no one
has yet dared to enter, for a 10-inch shell is lying on a table in the
middle. Nothing can be seen to touch it off but still people are suspicious. . . . Two of the magazines have not yet been opened, and we
shall have to be careful. . . . I noticed a heap of shell and cartridges
half buried in the sand, and I thought that prudence would dictate that
they should be touched . . . with a very long 10 foot pole.[5]

Even news reporters were impressed, but their accounts were
wisely restrained. "Inside the fort," dispatched a New York *Times*
man, "especially near the guns, and in the . . . streets, and . . .
places our men would walk, newly turned earth and other indications gave evidence of buried torpedoes. A guard was stationed
near all . . . suspicious spots. . . ."[6]

Louis E. Dawson, a private in a Maine regiment, drove the
ambulance carrying Generals George B. McClellan, William B.
Franklin, and Fitz-John Porter into the village. "Right in the
narrowest part" of the entrance, he saw one of the tell-tale red
flags and the top of a buried shell. McClellan chose this instant to
shout to an officer alongside the road: "Don't let our men take up
these. . . . Make the prisoners do it!" Dawson had stopped the
vehicle. In an unsoldierly fashion he objected when the general
ordered him to proceed, for he feared the torpedo would explode.
McClellan glared at the enlisted man. "Well, go ahead!" he
exclaimed, "I expect we'll all be blown to thunder together!"

Holding his breath, Dawson carefully guided the horses and
wagon through the gate. He did not know that McClellan had
scrutinized the shell and had noticed the absence of a tripwire.
Surmising that the wire had been removed, McClellan had used
the incident to impress his troops. It worked perfectly.[7]

An outraged McClellan vented his ire at mine warfare in a
telegram to his superiors: "The rebels have been guilty of the most
murderous and barbarous conduct in placing torpedoes within
the abandoned works near wells and springs; near flag staffs,
magazines, telegraph offices, in carpet-bags, barrels of flour,
&c . . . I shall make the prisoners remove them at their own
peril." To his wife he wrote: "It is the most murderous and
barbarous thing I ever heard of."[8] When Lincoln's Attorney

General, Edward Bates, read the dispatch, he made no secret of his indignation and noted it in his diary: "I hope he will put the prisoners of the highest rank foremost in this dangerous duty." [9]

The situation was far less serious than these statements indicate. First reports claimed that a great many persons had been killed or wounded, but contemporary accounts were often written under stress and embellished at the writer's discretion. Actually only a few of the buried shells were exploded, injuring perhaps three dozen men, some of whom died. One was a civilian—D. B. Lathrop, a telegraph operator—who set off a torpedo while planting a pole. [10]

The scheme had been originated by studious, fifty-nine-year-old Brigadier General Gabriel J. Rains, C.S.A., who commanded a brigade at Yorktown. His garrison totaled but twenty-five hundred men and faced the great host raised by McClellan, which Rains estimated at a hundred thousand. His position was in imminent danger of being overrun by sheer weight of numbers, and to compensate for the disparity, he put into effect a plan he had "invented" some twenty-two years before, during the Seminole War in Florida. In April, 1840, while a captain of the "old" army's Seventh Infantry, Rains was stationed at Fort King in the interior of the state. His outnumbered men were continually being ambushed by the Indians. Captain Rains made a land mine, using an artillery shell, and placed it under the clothing of a dead soldier near a pond used by the Seminoles. ("Booby traps," as they later became known, were frowned upon by professional soldiers, yet occasionally soldiers became as desperate as Rains and set out such hidden explosives.) A night or two later a loud booming awoke the garrison. Out they rushed, led by Captain Rains, expecting to find the earth littered with dead natives. Instead, there was only the carcass of an opossum. Indians hidden in the brush opened fire and drove the soldiers back into camp. Rains was wounded during the battle, but not seriously enough to prevent him from planting another torpedo. This one was never exploded, but it so unnerved patrolling soldiers that he was told to destroy it. Hunter Davidson, who was somewhat jealous of Rains's Civil War torpedo activities, related the event in a less than kindly manner: "The biter was bit, and the Indians caught him and peppered him with lead. He was daft on sensitive fuses, and his experiments were generally disastrous." This appraisal was far from true.[11]

Rains felt that his position at Yorktown was much like his position in Florida and required similar precautions. He had some of his men convert large artillery shells so that they could be exploded by pulling a string or by being trod upon, and these were buried among the fortifications. The soldiers understood too well and, in their enthusiasm, placed torpedoes "at spots I never saw." Undoubtedly they related this new sport to their comrades who in turn put more of the devices at other places in the town. In a number of instances, however, it seems that some of the cases were mere dummies put out to frighten the Yankees.

When the Confederates pulled out of Yorktown, the general sent guides to lead his men past the buried explosives: "We marched out and took our way . . . through Yorktown, under the guidance of a lieutenant . . . to enable us to avoid the torpedoes . . . planted in the road inside the works. . . . After leading us to the upper part of the town, the lieutenant, telling us we were out of any danger . . . disappeared." [12]

The Confederates fought a fierce delaying action at Williamsburg on May 5. During the withdrawal to Richmond the next day Rains's artillery was held up at "a place of mud slushes" in the road, several miles out of town. The guns could neither be wheeled to fight if attacked nor pulled out of the mud to continue the retreat. Hearing Union cannon shelling the road behind them to cover the advance of cavalry, General Rains ordered four 10-inch shells to be taken from a damaged wagon and buried in the road near a felled tree with strings attached to the fuses, "mainly to have a moral effect in checking the advance of the enemy . . . to save our sick," he explained. Federal cavalry ran upon the mines and set them off. More Yankees appeared, but hesitated, refusing to move further until they were allowed to examine the roadway. They were held up some three days. In the meantime, Rains freed his guns and got his ambulances safely to Richmond.[13]

The Confederates later maintained that the Union Army extravagantly overestimated both the number of torpedoes and the extent to which they were used on the Peninsula. No doubt the

Rebels themselves never knew just how many were actually laid since mines were used haphazardly and no records were kept. The salient points are not these debates, but what was actually accomplished. The Union advance was slowed—as General Mc-Clellan admitted. The retreating army had reached Williamsburg in time to form for a fierce delaying battle. Federal casualties were insignificant, but the troops learned such respect for and fear of the new weapon that the psychological effects became even more important than the physical damage.

When the Confederate high command found out about the mines, the matter underwent serious discussion. On May 11, 1862, General James Longstreet, who now commanded the Second Corps to which Rains was attached, forbade his laying additional torpedoes, because he did not recognize them as "a proper or effective method of war." (This opinion was shared by some of the Confederate enlisted men as well. Said one infantryman while being led through the minefield at Yorktown: "This is barbarism!") [14]

Rains refused to accept Longstreet's directive and appealed to the War Department. He drew an analogy between the use of pickets and mines: both were placed in advance of an army to warn it of the approach of the enemy, but when sentinels were used it often meant losing men. As to the claim that torpedoes were "not proper," he replied that "they are as much so as ambushcades, masked batteries and [underground gallery] mines." He claimed that McClellan was going to dig a mine under Redoubt No. 4 at Yorktown, and if the Yankees could plant explosives underground on a grand scale, why couldn't the Confederates put them out in small casings?

Along the way he picked up a valuable supporter, Major General D. H. Hill, who thought that all means of destroying their "brutal enemies" were lawful and proper. The final decision rested with Secretary of War George W. Randolph. Randolph approached the problem from a cold, rational, legalistic view. It all depended on the way in which the mines were to be used, he thought. For example, civilized warfare did not allow killing for its own sake, but only to achieve a definite military advantage. Thus, killing pickets with torpedoes was not to be permitted, but getting rid of generals by the same means was a different matter, for it deprived an army of its leaders. Mines could be placed in roads to delay pursuit, in front of the lines to repel attack, or in rivers and harbors to damage enemy warships.

Though these views vindicated Rains, Randolph found cause to chastise him in the grand old military manner for not giving way to the opinions of his superior, Longstreet, even if higher authority later overrode the superior. Tactfully, he suggested that Rains might accept a transfer to the James River defenses where there was no question as to the need for torpedoes.[15]

This suggestion was further strengthened by the request of the new commander of the Army of Northern Virginia. Robert E. Lee was trying to stop the Federals from advancing on Richmond. He was especially concerned about the Union Navy forcing its way up the James, where Confederate defenses were far from complete. The Navy had Maury working on electric mines but wanted something—such as the mechanical torpedoes—that could be made faster and put out in greater numbers. Lee called on Rains: "The enemy have upwards of one hundred vessels in the James River, and we think they are about making an advance . . . upon Richmond," he said. "If there is a man in the whole Southern Confederacy that can stop them you are the man. Will you undertake it?"

The answer, as well as the solution, to the controversy arising from the incident at Yorktown was obvious, and on June 18, 1862, Rains was put in charge of the "submarine defenses of the James and Appomattox Rivers." On the river bank opposite Drewry's Bluff he made his first water mines from gunpowder-filled lager beer kegs. Later he would boast that he was the first man to plant torpedoes in these waters—a claim that was not entirely valid. Maury had already set out his electric mines and had attacked Union ships in Hampton Roads twice with mechanical mines.[16]

Rains's appointment came but two days before the Navy selected Davidson to succeed Maury. The phrasing of these orders is worth examining, for there does not seem to have been a clear division of authority. Rains was assigned by the Army with instructions to call only upon military commands for assistance. Davidson was told to use his discretion as to "placing obstacles . . . of any character in the river" in addition to establishing the electric mines. Then too, Maury had written the Secretary of War when he was ready to mine the river. Such overlapping authority was bound to cause confusion, and on September 9, 1862, Rains turned over his James River command to the Navy. Henceforth the submarine bureau would care for the nautical defenses here, while Rains saw to it that other streams of the Confederacy were kept supplied with mines and torpedoes.[17]

CHAPTER IV

FIRST BLOOD

In Charleston, New Orleans, and elsewhere, the Confederates were laying the groundwork for the widespread use of torpedoes, but for the rest of 1862 the primary activity was confined to invention and testing. At this period both the weapons and the tactics were quite primitive. This was due as much to inexperience as to the lack of adequate materiel and facilities. An officer who had frequent opportunities to observe the development of mines wrote of these days:

Animated by an earnest desire to render their country some service, or perhaps attracted by the thought of the premium offered by the . . . Government for the capture or destruction of any vessel of the enemy, many an ingenious mind turned its attention to . . . inventing some *machine infernale*. . . .

The War Department and the chief engineer . . . were worse than opportuned by the . . . inventors, every one of whom demanded an examination of his plan or model. Such requests having to be granted for fear of possibly overlooking a perhaps really useful invention the attention of the examining committees would, naturally enough, often be called to the most absurd schemes. There were torpedo twin boats, propelled by rockets; diving apparatus by . . . which torpedoes might be attached to the bottom of the enemy's ship; balloons that were to ascend, and, when arrived just above the vessel . . . drop some kind of torpedo on the deck . . . ; rotation torpedo-rockets to be fired under water; submarine boats, with torpedoes attached to the spar; in fine, any variety of plans, and yet but a few, very few practical ones.

The great error which most of the inventors fell into was, that they aimed at accomplishing, all at once, too much in a field which to all of them was still an unexplored *terra incognita*. Complicatedness of the apparatus was the next consequence . . . which resulted in its utter failure in being tried. . . . Those torpedoes by which the heaviest losses were caused excelled in simplicity of construction and cheapness.[1]

Typical of the type of case noted by this engineer was that of E. H. Augumor, of Mobile, who developed a torpedo for which he made expansive claims. He succeeded in having his plans reviewed by Brigadier General Danville Leadbetter, who declined to offend the inventor by expressing his opinion. However, he was specific when writing to engineer headquarters: "I have no confidence in them."[2] The bureau referred the plan to Captain John M. Brooke of the Naval Office of Ordnance and Hydrography, who made short shrift of Augumor's idea, but in a diplomatic way.

Another case was that of West Beckwith of Richmond, who submitted several "improved plans of torpedoes for river and harbor defence." A persistent petitioner was a Mr. C. Williams who enlisted the support of Major J. A. Williams, an officer charged with torpedoes in the defense of Richmond and probably a kinsman. He forwarded plans and ideas for a torpedo and submarine apparatus. Sometimes he was not even graced with the dignity of an acknowledgment.[3]

These were but a few of hundreds of such incidents—a succession that continued unabated until the end of the war. A tiny percentage of the suggestions were worth considering, and from these emerged a few patterns of underwater obstacles that gave the enemy no end of trouble.

On February 13, 1862, a reconnaissance party from the Union fleet at newly captured Port Royal in South Carolina ran into just such weapons. Led by Lieutenant J. P. Bankhead, U.S.N., the unit's mission was to sound the mouth of Wright's River where it empties into the Savannah to find a channel that would lead them to the city of Savannah itself. Bobbing objects resembling empty tin cans were seen floating nearby but were ignored as flotsam. Then, Lieutenant J. G. Sproston of the *Seneca* rowed up and mentioned that he too had seen them and thought they might be buoys anchored to some of those new infernal machines. Bankhead paddled over and looked. Sproston could be right. There were five buoys in a line that stretched across the channel, which were visible only at low tide.

The following day three boats went out and dragged for the wires with grapnels. One object was pulled aboard. It was taken to the *Unadilla* for examination. Instead of the float, this was the torpedo itself, consisting of a tapered cylindrical body that was designed to lie on its side. Inside was an air chamber, powder, and—screwed into the top beside a lifting handle riveted to the center—a cannon friction primer attached to a wire. A pull of the wire would ignite the primer, exploding the torpedo. Commander John Rodgers regarded the mine with disgust and sent it ashore where it was destroyed by musket fire. "I suppose," wrote an onlooker, "these are what they meant when they said we would meet with a warm reception in coming to Savannah."[4]

That night the Federals were awakened about midnight by an explosion from the river. The *Susquehanna's* launch, towing an ammunition barge, tripped one of the torpedo wires as it passed the mouth of the river. Luckily, the mine blew up without doing any damage. However, Rodgers decided not to take any further chances, for he suspected that more torpedoes were set out upstream. The next morning Bankhead went out and sank the rest with muskets. "Now that we have discovered the plot," said a writer aboard one of the ships, "we shall be on the *qui vive* for them."[5]

A month later other torpedoes, their warheads attached to the tips of submerged chevaux-de-frise, were found in the Neuse River in North Carolina by ships moving up to attack New Bern. Small iron cylinders with barbs to affix them in the bottoms of wooden ships were found in the Tennessee.

It was to centralize the experimentation, training, and development of an efficient organization that the Confederate Congress in October, 1862, passed legislation relating to such activities. It authorized a Secret Service Corps, the Confederate States Submarine Battery Service, and the Torpedo Bureau. The first and last were military units, employing mechanical torpedoes made and used by soldiers or persons operating under authority of the Army or the engineers. The other was Davidson's naval organization which employed electrical mines. The instructions that had put Rains and Davidson in charge of their respective bureaus in June had been military orders only; these new directives had international implications, for they legalized torpedo warfare as far as the Confederacy was concerned and cloaked those engaged in it with legal protection and prisoner-of-war rights.

The new units were to be composed of "persons not otherwise liable to military duty . . . to be considered belonging to the Provisional Army . . . and entitled, when captured, to all the privileges of prisoners of war." Those who enlisted were sworn to secrecy and were granted a number of extra benefits including prize money based upon the value of enemy vessels they destroyed.[6] Included in the articles of enlistment was an agreement pledging the men "under no circumstances, now or hereafter, to make known to anyone not employed in this service, anything regarding the methods used for arranging or exploding the submarine batteries. . . ."[7]

Experiments were begun on other varieties of torpedoes and men were sent to England to recruit skilled technicians and obtain advanced data from laboratories set up there. The effect was the replacement of the crude and less perfect machines with ones "so certain and well devised" that the enemy held them in high esteem.[8]

The results were seen in the increasingly widespread use of improved mines. Electric torpedoes were initiated at Charleston; Rains's barrel mines were placed at a number of spots along the coast of South Carolina and in the channels of the Ogeechee near Fort McAllister, Georgia. Local commanders were ordered to try to force enemy ships to sail over them. More was heard of these explosives later, but the first success came in a sluggish waterway that few persons had any knowledge of—Mississippi's Yazoo River.[9]

At eight o'clock on the morning of December 12, 1862, a flotilla of light-draft gunboats, the U.S.S. *Pittsburg, Marmora, Signal,* and *Cairo,* and the ironclad ram *Queen of the West,* under the command of Captain Henry Walke, left their anchorage near Vicksburg, Mississippi, and began steaming slowly up the twisting, chocolate-brown, flood-risen Yazoo. The ships were on a twofold mission: to destroy enemy batteries on the thickly wooded shores and to destroy floating torpedoes, several of which had been spotted the day before.

Upriver, at Yazoo City, was a crude shipyard where at least three powerful ironclads were being built by the Confederates. There, too, were members of the Submarine Battery Service, Masters Zedekiah McDaniel and Francis M. Ewing, who had volunteered for the duty in August and had been given instructions at Vicksburg by Lieutenant Beverly Kennon. Kennon used a modification of Maury's mechanical torpedoes—powder-filled glass demijohns ignited by cannon friction primers when lanyards passing through watertight gutta-percha and plaster-of-paris plugs were pulled. The action can be likened to the striking of a match in that a treated wire was pulled across a rough surface.[10]

The fleet proceeded peacefully along the Yazoo until the *Marmora,* leading, came under hot musket fire when she attempted to negotiate a tight bend. She was backing and filling for the turn when the *Cairo* (a 512-ton turtle-like ironclad designed by S. M. Pook and built in St. Louis by James Eads) joined her. The crews manned the 8-inch rifles and 32-pounder smoothbores, searching the banks for the enemy. The *Cairo's* skipper, Lieutenant Commander Thomas O. Selfridge, Jr., expressed surprise when the *Marmora* opened up on a block of wood in the water, and ordered her to stop firing and send out a boat instead. The lead ship, not hearing the command, did neither, and the *Cairo* put out her own boats. The object was recovered and declared to be part of an exploded torpedo.

Ensign Walter E. H. Fentress, in a cutter belatedly launched from the *Marmora,* discovered a line in the water. Believing it to be a torpedo lanyard, he severed it with his cutlass. Immediately, a mine bobbed to the surface. Another line was located and Selfridge told him to cut that one as well. Instead, Fentress pulled the line and towed the torpedo ashore. Fortune (which often has a high regard for impetuosity) was with him, for the mine did not explode. While he was breaking up the torpedo, Fentress heard an explosion, looked up and saw a ship's anchor flying through the air. It was the *Cairo*'s.[11]

While attention had been riveted on the boats, the *Cairo* had been allowed to drift toward the shore and had to be backed into the stream. Both she and the *Marmora* then commenced to move ahead slowly. They had traveled less than half a length when, as Selfridge wrote:

Two sudden explosions in quick succession occurred, one close to my port quarter, the other apparently under my port bow—the latter so severe as to raise the guns under it some distance from the deck.

She commenced to fill so rapidly that in two or three minutes the water was over her forecastle. I shoved her immediately for the bank, but a few yards distant; got out a hawser to a tree, hoping to keep her from sliding off into deep water. The pumps, steam and hand, were . . . manned, and everything done that could be.

Her whole frame was so completely shattered that I found . . . that nothing more could be effected than to move the sick and the arms. The *Cairo* sunk . . . minutes after the explosion, going totally out of sight, except the top of the chimneys. . . . Though some half a dozen men were injured, no lives were lost. . . .

There was perfect discipline. . . .[12]

The first Union reports had it that the *Cairo* was a victim of the much-discussed "galvanic torpedoes," and even experts became confused as to the particular type of explosives used. Actually, the ship fell prey to a pair of five-gallon wicker-covered demijohns in wooden boxes, connected by twenty feet of wire and suspended below the surface. The ship struck the wire, pulling the bottles onto her sides and igniting both friction primer fuses. Since the explosions occurred under the bow and on the port quarter, it appears that the *Cairo* almost missed the torpedoes, and struck the wire very near the torpedo at the right side of the channel. The bottle was held against the bow by the ship's forward motion; this in turn pulled in the second mine, which struck the port side and exploded. The torpedoes were held in place by wires stretched from the shore as well, a fact that misled the Federals into assuming that galvanic batteries were located ashore.

Watching the operation from a rifle pit were Ewing, McDaniel, and Lieutenant Isaac M. Brown. The latter had set out Maury's electric mines near Columbus, and he became famous in history for commanding the Confederate ram *Arkansas*. He said that he was torn by conflicting emotions when the *Cairo* sank, "much as a schoolboy . . . whose practical joke has taken a more serious shape than he expected." [13]

Later, McDaniel and Ewing submitted a claim to their government for a reward for the sinking, "on the ground that the torpedo . . . was invented by them." The application was refused in office after office, but the persistent pair kept appealing until Congress nearly passed a resolution in their favor. The matter reached the desk of President Davis. A thorough investigation initiated by the President refuted several of their allegations, but the matter hinged on the question of the "originality" of the torpedo. It was discovered that the weapon was not original with them at all, and even (for argument's sake) if it were, Ewing and McDaniel had violated a statute that required inventors to provide the government with plans and allow it to "enjoy freely its reserved rights of using the invention in its own service." Finally, the pair were on government service at the time, and were doing no more than carrying out their required duties. After all, the torpedo had been made in a government shop, with government materials, and by persons in government employ. If it had failed to explode would they have billed the Confederate States for their trouble? [14]

After rescuing the crew of the *Cairo*, the *Queen of the West* (heaviest ship in the flotilla) came up and pulled away the stacks to keep the Confederates from finding the wreck. Other ships shelled the banks to drive off Rebel sharpshooters and torpedo operators while small boats scoured the water for additional mines. A dozen were located and destroyed, as was Brown's manufactory, complete with powder and cases, which was located nearby. [15]

Instead of keeping Union Admiral David D. Porter out of the Yazoo, the sinking seemed to spur him into greater activity. He laid out plans for another venture upriver not very long afterwards. The orders reveal the respect for torpedoes he had tried to instill in his captains. "The brightest lookout must be kept," he wrote. "Have plenty of rowboats to go ahead . . . with drags and searches of all kinds." Cutters were to scrape the edges of the banks with grapnels to snag wires while the ships dragged the channel. Once a floating object was found, it was to be approached with caution and, if it was a torpedo, hauled ashore with a long line so that it could be destroyed. The sailors were furnished information so that the officers could "keep a diagram of the river and the manner in which these torpedoes are laid, for the rebels have a regular . . . system which we can avoid when we know it." [16]

The gunboats *Baron De Kalb* and *Signal*, the *Queen of the West*, and a tug went up the Yazoo again on the twenty-third of December, shelling the woods along the way. Confederates hidden in the thickets shot back, harassing the ships. The next day, near the *Cairo* wreck, the *Queen* ran over a torpedo wire which a lookout had seen but which the ram could not avoid. Officers and crew froze in expectation of an explosion. None occurred. Boats were lowered and the ironclad waited for them to explore the water, but riflemen began peppering her again, driving the seamen in the rowboats back to their mother ships. The gunboats took up the firing and both sides blasted away for the duration of Christmas Eve.

On December 27 army troops under Brigadier General Frederick Steele moved up along the bank to enfilade the levee where a Confederate battery was mounted. Meanwhile the fleet proceeded cautiously behind cutters which were pulling up torpedoes. The *Queen of the West* sailed ahead of the pack and promptly ran aground. Luckily she managed to pull herself free before the Rebels could wheel their artillery into position, but this nearly proved to be the end of the expedition. The ships stopped off the Benson Blake Plantation, where sailors destroyed buildings by detonating captured torpedoes which they had brought ashore.

Admiral Porter renewed the push upstream on the last day of 1862. The *Lioness* was fitted with a huge 65-foot torpedo rake, an invention of Colonel Charles R. Ellet. It was supposed to hook the mines and pull them to the surface. This was the first appearance of a contrivance that would become standard with the Navy during the Civil War. Made of logs from which dangled scores of grappling hooks, the rake was boot-jack shaped and was widely copied in the service. It was quite possible that one of the torpedoes caught by the contrivance might explode under the ship instead of under the rake, but this worried neither Porter nor Ellet. "If we lose her," wrote the former of the *Lioness*, "it does not matter much." Ellet was more hopeful, and was of the belief that the mines would detonate so far in front of the ship that she would not be destroyed. What her crew thought is not recorded, but it is not difficult to surmise. To their relief, a dense fog followed by heavy rain caused the admiral to cancel the whole expedition.[17]

Porter summed up his experiences rather engagingly a few days later:

We have had lively times up the Yazoo. Imagine the Yazoo becoming the theater of war! We waded through 16 miles of torpedoes to get at the forts (seven in number), but when we got that far the fire on the boats from the riflemen in pits dug for miles along the river and from the batteries became very annoying, and that gallant fellow [Lt. Commander William] Gwin thought he could check them, which he did until he was knocked over [by cannonfire] with the most fearful wound I ever saw. He could not advance, the torpedoes popping up ahead as thick as mushrooms, and we have had pretty good evidence of their power to do mischief. I never saw more daring displayed than by the brave fellows who did the work. The forts are powerful works, out of the reach of ships, and on high hills, plunging their shot through the upper deck, and the river so narrow that only one vessel could engage them until the torpedoes could all be removed.[18]

Some time later Isaac Brown told a humorous story which was supposed to demonstrate how certain other Federals felt about these weapons. When a Confederate chanced to board one of the Union gunboats under a flag of truce, he met a former messmate. As the two joked and bantered, the Southerner asked, "Tom, why don't you go up and clean out the Yazoo?"

"I would as soon think of going to Hell at once," was the reply, "for Brown has got the river chock-full of torpedoes!"[19]

CHAPTER V

SPREADING SUCCESS

$\mathbf{A}$s the news of this grand success spread in the South it encouraged interested Confederates into redoubled efforts at invention and production. Gabriel Rains concentrated his energies on a sensitive primer designed "to explode from the slightest pressure," eventually stabilized at seven pounds. The "Rains Fuse" was effective either on land or in water, but, unlike certain others, it was intended solely for static defense and when set and armed was as dangerous to friend as to foe.

The exact composition of the fuse remained a closely guarded secret until after the war. Some sources say that only the inventor knew it and that he neither wrote it down nor allowed anyone else to supervise the construction of the primers. As revealed later, the formula was as follows: 50 per cent potassium chlorate, 30 per cent sulphuret of antimony, and 20 per cent pulverized glass. These primers were activated when a thin copper protective cap was crushed, causing the composition to detonate and thereby igniting a fuse that communicated with the powder. The quick match fuse was made of gunpowder dissolved in alcohol.[1]

Two types of water mines came from Rains's shop: the famous "keg" torpedo and the "shell" or "frame" torpedo. The first was efficient in its simplicity. Small wooden kegs—at first those used for lager beer and later some specially built for the purpose— were modified by the addition of wood cones on the ends, a coat of pitch was applied inside and outside, and a sensitive fuse was inserted in the bung. Weights attached to the bottom floated the fuse-side up. Cheap, easy and quick to make, and deadly efficient, these caused more destruction than any other torpedo.[2]

It was with the frame torpedo that Rains scored first. This apparatus was the marrying of his fuse with hollow, 400-pound, 15-inch artillery shells cast with thin upper surfaces to direct the explosive force upwards. The fuses were screwed into an opening at the nose, and the shells were attached to the heads of a row of heavy timbers. The other ends of the timbers were tied into large wood frames or "cribs" anchored to the bottom of the waterway, often with great boulders and stones. The timbers and shells (each containing twenty-seven pounds of powder) were raised until the deadly fuses were poised just beneath the surface, ready to be crushed against the bottom of a passing ship. When placed in sections across rivers and bays this was an extremely effective barrier, so flexible that its depth could be regulated simply by adding or removing weights. U.S. Navy specialists assigned to study the devices reported that "our gunboats never . . . attempted to force a passage through a channel known to have been defended by . . . them."[3]

Frame torpedoes were soon planted in waterways all over the South. When General P. G. T. Beauregard learned about them he took steps to see that they, like a host of other new weapons, were made available to the areas under his command. In early 1863 none of his commands was in greater danger than Fort McAllister, which guarded the approaches to the Ogeechee River near Savannah. The Yankees were giving much attention to this fort for two reasons: they wanted to control the river itself and to capture or destroy the famous blockade runner *Nashville* which was protected by the fort's cannon. This ship—the very first to come from England with war supplies while flying the Stars and Bars—had been bottled up for three months. Her position, according to a contemporary authority, was "unassailable." The fort, he said, was "strong and well-constructed . . . so placed as to enfilade the narrow and difficult channel for a mile below. The river had been staked opposite the fort, and a line of torpedoes had been planted at intervals lower down. . . . Above . . . lay the *Nashville,* ready to dash out at the first sign. . . ."⁴

Three Union gunboats had been stationed at the river's mouth to prevent the *Nashville* from making a bid for freedom, but they had so few guns that they could only remain outside and watch. In January, 1863, Commander John L. Worden (famous for having directed the *Monitor* in her battle with the *Virginia*) was dispatched to the scene with the brand-new, turreted monitor *Montauk* which Admiral Samuel F. du Pont was anxious to pit against land-based artillery. Worden's squadron was composed of six vessels: the U.S.S. *Seneca, Wissahickon, Dawn, Williams, James Adger* (which had towed the monitor down) and the *Montauk,* all of which anchored well out of range of the fort on the twenty-sixth while a pair of boats reconnoitered. Lieutenant Commander John L. Davis reported that a line of piles crossed the stream below the fort and he thought some of them bore frame torpedoes. At seven the next morning the fleet, carefully avoiding the suspicious obstacles, moved in and opened fire. The battle between ships and fort raged until noon, ending in a draw. After ammunition and stores had been replenished, the vessels ventured in again on February 1 and repeated their previous maneuver with similar results. There the stalemate remained, tacitly acknowledged by both sides.

At dusk of February 27 the *Nashville* was seen moving as if getting ready to make a run for the open sea. Worden signaled his fleet to attack at daylight. During her maneuvers the *Nashville* somehow contrived to ground herself, and at dawn her helpless condition was noticed by the alert Federals. Cautiously, they formed column and moved out of the roads into the river mouth, taking positions with deliberate care, even though they had come under the fire of Fort McAllister. Geysers climbed high from the water where shells struck, but none of the ships was hit. The *Nashville* joined the cannonade. Between the opposing vessels was half a mile of green marsh over which they fired. Shot after shot smashed into the immobile Confederate ship, and in less than twenty minutes she was reduced to a shambles. She burned fiercely and exploded with a great flash and roar when the fire reached her magazine.

Their mission accomplished, the Union squadron broke off and withdrew, carefully keeping in the same tracks they had used when entering the river. At 9:35 A.M., a few yards above a creek known as Harvey's Cut, the *Montauk,* still under fire from the fort, shuddered, raised in the water, and slewed around violently knocking her crewmen off their feet. There followed what was described as "a violent, sudden and seemingly double explosion." The first thought of the crew (and the gunners in the fort as well) was that one of the Confederate guns had scored a direct hit. Water poured through a rupture in the *Montauk's* iron skin. Anxious questions from the bridge were answered by the engineer, who thought she could be held afloat by the pumps, but in a few minutes it became apparent that all efforts were futile. Half a foot of water sloshed about on the engine room floor, playing at the bases of hot furnaces as sweating stokers waded through it to heave coal into the fires. Reversing his decision, the engineer shouted up the voice tube that, to save the ship, he advised she be run aground as quickly as possible.

The *Montauk* was steered onto a nearby mudbank, and at low tide an inspection was made of the hull. Some of the bottom plates had been forced upward three and a half inches in a three-by-five-foot area; a ten-foot crack zigzagged down her side—all evidence of tremendous strain. A spare piece of boiler iron was patched over the rip, and, with her pumps running, the monitor managed to stay afloat during the tow to Port Royal, South Carolina.[5]

At this same time, in the Mississippi River near Port Hudson, Louisiana, the Confederates attempted to disable another Union warship. She was the "tinclad" gunboat (a river steamer with a thin iron coating) *Essex.* During a patrol off Profit Island her skipper, Commander C. H. B. Caldwell, saw a buoy through his binoculars. Between that buoy and the shore was another and, not far from there, a semisubmerged barrel. Caldwell wanted to stop and examine them, but as the *Essex* was chopping along at top speed with a strong wind behind her, he could not slow down in time. Later, going back down the river, Caldwell sent his crew to quarters while a pair of cutters rowed over to the buoys. They found a wire running from the floats to the shore and, hanging below the surface, a pair of cylindrical iron torpedoes, three feet long and one foot in diameter, which were "finished in a most workmanlike manner." From each a wire led to the shore. One was taken aboard and studied by Caldwell, who wisely sent it to shore for disassembling. Once ashore, the torpedo was placed in a hole dug in the levee and the wire was pulled from a rowboat. The explosion tore away large chunks of earth and scattered iron fragments over a wide area—one two-pound chunk landed on the *Essex*, three hundred yards away. Though Caldwell did not know it, this was one of the Brown-McDaniel-Ewing-Kennon torpedoes.

Caldwell found another torpedo a few days later, on the night of February 27, 1863. This was the first of a new type—the mechanically operated mine. It was a clock torpedo containing a tin cover, 114 pounds of cannon powder, and a clockwork device connected to two hammers poised over a pair of pistol-sized percussion caps. The mechanism was running perfectly when the mine was captured, but Caldwell's time had not run out.[6]

Nor had the luck of the 1,929-ton screw sloop *Richmond*. On the night of March 14, 1863, as part of Admiral David G. Farragut's squadron that was running up the Mississippi past the Confederate batteries of Port Hudson, the *Richmond* detonated a torpedo. The weapon exploded almost beneath her stern, throwing up a column of water thirty feet high, shaking the ship, bursting windows, and knocking parts of a 9-inch gun over the side.[7]

On January 24 of that year the Confederates had suffered a "casualty." Rains was working with his primers in his Richmond laboratory when one exploded in his right hand. The injury was to the thumb and forefinger, and an acquaintance later wrote that "he was scarcely able to sign his name to official documents today" as a result.[8] It was not painful enough, however, to prevent Rains's working on a book about his invention. On the following day he revealed some of the book's contents. Technical details he omitted, explaining only his reasons for engaging in warfare with the torpedo.

"He says," wrote John B. Jones, a War Department clerk, "he would not use such a weapon in ordinary warfare; but has no scruples in resorting to any means of defense against an army of Abolitionists, invading our country for the purpose, avowed, of extermination." Gabriel Rains appears to be one of the few Confederates who used torpedoes with such bitterness. This may have been due partially to his belief that the shelling of his home town (New Bern, North Carolina) by the Federals was done without advance warning to the population. Other men seem to have sought either adventure, pure military advantage, or financial reward, but to him it was a very personal thing.[9]

It is unfortunate that his book (actually a long manuscript report, complete with diagrams) did not survive. From all accounts, it was carefully done and would provide an interesting contrast with studies compiled by Union officers after the war. It was submitted to President Davis, both to acquaint him with progress in development and to enlist his support in the adoption of the Rains designs. There is no question that it was effective. "It caused him to enter with zest into the schemes," said Rains of Davis, and the Chief Executive himself admitted to sending extracts to generals in the field. Only a single complete copy was made and this was accompanied by a request that it be printed and provided to all departments. The President wisely refused because "no printed paper could be kept secret." He felt that this weapon "would be deprived of a greater part of its value if its peculiarities were known to the enemy." Thus the most widely used "new" weapon of the Civil War remained a closely guarded military secret.[10]

Rains's report was even more persuasive than he dared imagine. In fact, it so impressed Davis that almost before he knew it, the inventor found himself the subject of Special Orders No. 124 of May 25, 1863, sending him into the field. "The President," said a letter with the orders, "has confidence in his [Rains's] inventions and is desirous that they should be employed on both land and river . . . at Vicksburg and its vicinity." Commanding the armies in the West was General Joseph E. Johnston, an avowed opponent of the use of mines, who was warned that the weapons had been "approved and recognized" by the government.[11]

Thunderstruck at the suddenness of the command, Rains appealed to the President, stressing that he could not possibly prepare significant numbers of torpedoes soon enough to change the tide of a battle already in progress. Davis cut him off on June 3 with a curt: "I had expected you would have secured haste in your movements, . . . yet, . . . you have not started." [12] The plan was to mine the Mississippi and the roads that the enemy would take when they attacked Vicksburg.

General Rains left the capital as soon as he could. Instead of assembling his explosives in Richmond, he decided the best thing to do was to take the paraphernalia he needed along with him inside a specially built wooden chest which had cost him twenty-seven dollars. The mines would be assembled at his destination. Inside the chest were "sensitive primers & chemicals for construction." When Rains reached Augusta, Georgia, he requisitioned other material from his brother, Lieutenant Colonel George W. Rains, brilliant superintendent of the great government-owned powder mill. At Jackson, Mississippi, he rented a room in which he set up a laboratory to make fuses. Few additional chemicals were needed and the bill from Dove & Company of Jackson was for only $4.50. If only a few Yankees were delayed by the torpedoes, Rains's trip would have been a good bargain for the Confederacy, for the complete statement of expenses as submitted for collection was but $397.[13]

Very little was achieved although reports were current that when Johnston retreated to Jackson, "hundreds of the enemy . . . were killed and wounded" by the torpedoes. It was hoped "this invention may become a terror to all invading." [14] What actually happened was, of course, quite different. Rains's protest had consumed valuable time and the Federals did not wait for him. "There could scarcely have been presented a better opportunity for their use than . . . the heavy column marching against Jackson," President Davis ruminated bitterly, "and the enemy would have been taken at a great disadvantage if our troops had met them midway between Jackson and Clinton." [15]

From a broad view, however, there was some consolation to be gained from the resulting appearance of torpedoes that were the products of other inventors, among whom were two Texans, E. C. Singer (a gunsmith relative of the sewing machine manufacturer) and Dr. J. R. Fretwell. Their approach to the problem of ignition was different from Rains's and depended upon the action of a strong spring. The mine consisted of a floating tin cone two-thirds full of powder. An iron rod with a plunger and the spring extended through the case and an equal length below it. The weight of a saucerlike iron plate falling from the deck of the cone yanked a pin safety, thus releasing the spring-driven plunger which smashed a percussion cap inside the body of the torpedo. It was the utter simplicity of the mechanism that swayed an examining board in July, 1863:

> The lock is simple, strong and not liable at any time to be out of order; and . . . the caps . . . are not likely to be affected by moisture. . . . By the peculiar and excellent arrangement of the lock . . . the certainty of explosion is almost absolute. One great advantage this . . . possesses over many others is, that its explosion does not depend upon the judgment of any individual; that it . . . cannot readily be ascertained by an enemy's vessels.
> We are so well satisfied . . . that we recommend the engineer's department . . . have some . . . placed at an early date in some of the river approaches to Richmond.[16]

Soon clusters of the dull-gray containers were anchored in rivers all over the South. In fact, the Fretwell-Singer torpedo appears to have been used more frequently and in greater numbers than any other model with the exception of Rains's kegs. Some, made in Yazoo City by the inventors, were placed in the muddy Yazoo River that same July by the indomitable Isaac Brown, lest the Yankees try to ascend that hotly contested stream again.[17] It was as if Brown had been reading Admiral David D. Porter's mail, for on July 11 Porter received a note on that very subject from Major General U. S. Grant, commanding the armies around Vicksburg: "Will it not be well," Grant asked, "to send up a fleet of gunboats and some troops and nip in the bud any attempt to concentrate a force there?"[18]

This came at an appropriate time for Porter, because he was able to divert some forces from Port Hudson, which had just surrendered. Several ships set out, including the armored gunboats *Baron De Kalb* (formerly the *St. Louis*), *New National*, *Kenwood*, and *Signal*, under the command of Lieutenant Commander John G. Walker, plus five thousand soldiers under Major General Francis J. Herron.

As they crawled cautiously along on July 13, 1863, the *De Kalb*, a 512-ton former riverboat mounting eighteen guns of various sizes, engaged a battery just below Yazoo City, across from the navy yard. The Confederate fire grew so accurate and rapid that the ship was forced to fall downstream while Herron landed his soldiers. Together, ships and men made a combined attack and captured the city, but not without loss to the fleet. While supporting the assault the *De Kalb* was shaken by a disturbance under her bow. Water cascaded through a large hole made by a Singer torpedo, and the ship settled quickly. A few minutes later a second torpedo blew up under the stern, rending the fragile hull once more. Within fifteen minutes the remains of the *Baron De Kalb* lay on the bottom of the river a total wreck. All of her crew were rescued.[19]

The Federals exacted a penalty from the citizens of Yazoo City, seizing three thousand bales of cotton. It was said to have been easily the equivalent of the value of the sunken gunboat. Wrote David D. Porter: "But for the blowing up of the *Baron De Kalb*, it would have been a good move." He reasoned later that "this is a fortune of war—to achieve anything risks must be run. . . ."[20]

With the capture of Port Hudson and Vicksburg, the United States had carved the Confederacy nearly in half by slicing it down the Mississippi. It was imperative that the Confederates rejoin, but to do so, they needed to regain control of the river. They never really tried to do this, choosing instead to remain on the defensive and battle the Union ships whenever they penetrated the Mississippi's tributaries. Such measures were forced upon them for several reasons, one being the ability of the U.S. Navy to maintain the offensive along the whole river. Another reason was the Union's preponderance of ships which enabled them to send out frequent patrols and detect any sizable force of Rebels. Much of the credit for this state of affairs must go to energetic, able David Dixon Porter.

Though the *Cairo*—the first ship ever sunk in combat by a torpedo—was lost in these waters, mine warfare never was established as it had been to the east. Torpedo men found it more difficult to make their "plantations" in the swift waters which bore trees and other debris to uproot or explode the torpedoes. Frequent Union patrols made the setting out of torpedoes from small boats especially hazardous; consequently, most of the work was done from the banks. However, the shores too were under Yankee domination most of the time, and torpedo-laying parties were forced to work by stealth. Naturally, such conditions ruled out the establishment of permanent mine stations by the Submarine Battery Service and made it next to impossible to operate torpedo boats of the type being used at Charleston. This left any torpedo-doing that might be done up to Rains's Torpedo Bureau or the Secret Service Corps. These daring men made the best of the few opportunities presented them—at times even surprising their own most sanguine hopes.

Enough tin Singer torpedoes were assembled at Port Lavaca, Texas, to present a serious obstacle for the Federals who attacked the southern Texas coast in the fall of 1863. Guarding Matagorda Bay was Fort Esperanza at Pass Cavallo, the entrance from the Gulf of Mexico, which was brought under siege November 27. Captain D. Bradbury, who was in charge of the Confederate torpedo operations, began planting his machines in May, 1863. Not only was he familiar with these waterways from having lived there but he probably built the torpedoes himself, for E. C. Singer had been employed in Bradbury's Foundry in Port Lavaca as a gunsmith. Eighteen mines were anchored in the channel between the fort and the bar at Pass Cavallo, but nothing much came of the effort. For one thing the water was too deep (thirty feet) and a strong current tore some of them away. Apparently they were seen by the Union ships, for none ventured into the area.

Just before the Federals attacked Fort Esperanza, Bradbury set out two dozen torpedoes, all he had left, in trenches around the work.[21] They were described by Captain T. L. Baker, U.S.A., after the capture of the fort:

I recovered and opened 13 of one kind, and 3 of another; the first were applied to land defence, and were taken from the dead angles at each salient and from the West curtain face. . . . The shape of these, was a cylindrical vessel of from 13″ to 21″ in diameter, and from 15″ to 24″ in depth. These cases contained on an average 35 to 40 lbs of fine diamond-grained powder. The latter were intended for submarine defence, and were the same pattern, with the exception of an air-vessel attached. The composition of all was tin; the latter with a prolongation of the standard, to admit of passing the striking-rod through the air vessel.
I discovered the position of the land obstructions by passing along

the superior slope of the work and noticing at first peculiar lines on the exterior slope, in such positions as would not identify them as having been used for profiles. On examination of the decayed grass lines, I found, by removing the grass, wooden troughs or boxes, through which raw hide lines had been run. In following these up, I invariably came to a torpedo; these I lifted, after first cutting the lines and then seizing the brass pin and doubling it while it was in position.

The land torpedoes were invariably laid upon their sides and were braced on their resistance side by a log of wood, or else stakes driven into the ground; and to protect the machinery from clogging with sand, they were covered with boards, strong enough to withstand a fair weight of earth. Around the exterior of the case were bedded quantities of small stones, which, in the event of an explosion, would act as splinters, or makeshift projectiles, and in rainy weather would serve as drain-ways to keep the torpedo in more perfect condition than if it were fairly in contact with the earth.

The explosion-lines or lanyards reached entirely inside the Fort, so that in case of an assault, they could be pulled at the proper time, (when the . . . angles were crowded,) with perfect safety to the garrison.[22]

As the Rebels were forced away from the coast, Bradbury took care to distribute his mines behind him. Four were put in a narrow channel near the mouth of the Guadalupe River to the southwest in such a way that no vessel could come from Aransas Pass without exploding them. More, including newer and more powerful models, were set in an artificial channel that ran near Port Lavaca. No Union ships suffered damage from them, however.[23]

Meanwhile, trained operators were being sent to the West from the torpedo factories along the Atlantic seaboard. One group had its headquarters in Mobile and included in its list of personnel Singer and Fretwell, both of whom remained in the workshop "to superintend the construction and management." Besides torpedoes, they hoped to build small ironclad torpedo boats.

They laid their plans at Shreveport, Louisiana. The first and obvious maneuver was to block the Red River to protect the ironclad *Missouri*. Below the town of Grand Ecore some thirty torpedoes were put out in March, 1864, firmly anchored beneath the swirling yellow water. The river was narrow and there was no other channel. Any intruding vessels would surely run over them.

Soon afterwards a strong United States force carrying troops moved up the river, intent on capturing Alexandria, Louisiana, as part of a campaign designed to drive the Rebels farther into

Texas. This fleet was called "the most formidable force . . . ever collected in Western waters." There were thirteen gunboats and seven lighter vessels.[24] At Fort De Russy, on the way north, the soldiers landed and captured the fort, then took Alexandria and pushed on toward Shreveport. However, the army was weakened at the battles of Sabine Cross Roads and Pleasant Hill below Shreveport on April 8 and 9, and began to fall back downriver, harried by the Confederates. Unseasonably low water made the navy's retreat more difficult and slowed their movement.

On April 14, about a mile below Grand Ecore, the ironclad *Eastport* (a former Confederate ship of 700 tons) grounded. It took all day to get her free, but she made some eight miles the next morning, preceded by cutters which destroyed two torpedoes. The *Eastport* was moving along carefully, a leadsman in the bow sounding the bottom with his line, her wheels idling, when she suffered what her skipper regarded as a "shock." The man up front was nearly thrown overboard, but Lieutenant Commander S. L. Phelps hardly felt it in his cabin. The hold began to fill slowly and, to save the ship, Phelps ran her aground. His efforts failed, however. The *Eastport* sank in five hours; a torpedo had exploded under her bow.

When Admiral Porter was informed, he sent two steam pump boats, *Champion No. 3* and *Champion No. 5*, to raise her. They lowered the water in the hold so that she could get off the shoal. To lighten her, her guns were removed as were the "heavy articles." She grounded eight times while being taken forty miles downstream. When she grounded again, Confederates began firing at ships around her, and she would not budge. Porter decided to destroy the vessel; a ton of powder was placed in the hull and, on April 26, the *Eastport* was blown up.[25]

CHAPTER VI

FROM CHARLESTON TO THE JAMES

In the fall of 1862, when General P. G. T. Beauregard returned as commander of the defenses at Charleston, he received a hero's welcome. Beauregard was admired in that city for having bloodlessly reduced Fort Sumter the previous year. Before leaving he had painstakingly outlined and supervised the construction of several lines of defense to protect Charleston and its harbor from the Union attacks that he knew were imminent. Upon his transfer to Virginia these defenses had been partially completed. Instead of finding them in readiness in 1862 as he expected, Beauregard was appalled to learn that some had progressed no further than when he had left, others were untouched, and, worst of all, his successor, Major General John C. Pemberton, had completely shifted some lines, passively permitting the enemy to occupy vital outlying areas.

With characteristic vigor the "Little Creole" set about making up for lost time. Troops, civilians, and slaves labored from sun to sun digging embrasures, moving cannon, cutting fields of fire, and transporting supplies and ammunition. In the meantime, dark ships of the Federal Navy became increasingly numerous off the harbor, tightening their blockade.

At this juncture, Beauregard turned his attention to defense of the waters as well as the land surrounding it. A boom of logs and rope was built across the main channel, which lay between Forts Sumter and Moultrie, gunboats and ironclads were stationed at salient positions, and the heaviest guns he had were trained on the channel.

Successful experiments with electric torpedoes had convinced some department commanders that these weapons would be desirable in their areas of defense, and by 1863 a number of officers who had trained with Maury and Davidson had been sent to help them. One of these men was Engineer Captain M. Martin Gray, a tall, dark, forty-two-year-old native of Delaware who took charge at Charleston. There, the "electrical department," run by a civilian, M. J. Waldron, had been in the process of conducting tests of its own. After Gray's arrival, this department concentrated on small electric torpedoes connected by a single cable instead of on large tanks. In their first effort Waldron and Lieutenant Charles G. De Lisle intended to add torpedoes to the boom across the main channel. Their weapons were fifty small "tin-iron" containers on a 3,500-foot insulated copper cable which had been purchased in Nassau for ten dollars a foot. The devices were a departure from the other kinds of electric torpedoes in that they were self-detonating: each mine had spreading armatures that completed a circuit and set off the explosion on contact with an iron-hulled ship.

On paper the idea looked sound enough, but in practice a number of flaws developed: first, the inventors learned that the system required more insulated cable than they had on hand; next, they found that the mine containers collapsed in only eight feet of water when they were weighted; and, last, they found that the tides would fray the cable by dragging it across the bottom unless it was securely anchored. In the end, they settled for mechanical torpedoes made by Captain Francis D. Lee.

An iron boiler, eighteen feet long and three feet in diameter, was equipped for electrical detonation. The fuse was made of two carbon "pencils" wrapped with wire in a 10-inch glass phial, their points a fraction of an inch apart. The tube was filled with fulminate of mercury and sealed, then placed in the center of the boiler into which was poured three thousand pounds of gunpowder. Cables connected the fuse with a battery ashore. The mine was secured by iron straps to a large raft, slung between a pair of barges, and floated out beyond Sumter where it was anchored to the bottom. The towing steamer *Chesterfield* ran aground during the operation but was refloated without damage. On shore, Captain Langdon Cheves waited, ready to throw his electric key the instant the Signal Corps at Batteries Gregg and Wagner on either side of him notified him that a Yankee man-of-war was positioned over the torpedo.

The plan had been set in motion when the Signal Corps learned of a pending Federal attack which came on April 7, 1863. The Union monitors, led by the new armored ship *New Ironsides*, sailed into the harbor and dueled with shore batteries, but they did little damage and suffered a great deal themselves, including the loss of the *Keokuk*. The *Weehawken* was pushing a 50-foot "devil," similar to the rake used by Porter on the Yazoo, which was designed to grapple and explode torpedoes. Confidently, she ventured near the row of percussion torpedoes (some of which were dummies) which the Confederates had strung across the channel. Shortly after 3 P.M. she succeeded in exploding a torpedo, lifted slightly, but remained sound. Her greatest injury was above the waterline, for she was hit fifty-three times in forty minutes by shells from the Confederate batteries.

Meanwhile the *New Ironsides* dropped anchor farther out while her crew served the guns. Jubilantly, the Confederate signal operator at Battery Gregg waved his flag, and was followed by the one at Wagner. Cheves, a companion wrote, "could not have placed [her] more directly over [the torpedo] if he had been allowed to." The contact was closed, and all eyes were glued on the Federal ship.

Nothing happened.

The switch was thrown again, and again, and still once more. The wire leading from the battery was examined—everything appeared to be in order. One more attempt was made, but as a disenchanted officer reported, "the confounded thing, as usual, would not go off."

Untouched and unaware of the enormous danger in which she had lain, the great black *Ironsides* leisurely weighed her anchor and sailed out of the harbor with the rest of the fleet. Later, the mechanics of the mine were thoroughly studied, and the cable was given an inch-by-inch examination. Where it crossed the beach the cable was broken, a set of wagon tracks showing the cause. The responsible party was never discovered, but suspicions were rife. Some felt Langdon Cheves had done it purposely; others blamed Captain Gray. One man reported that Lieutenant J. D. Blake was arrested for having said he hoped the torpedo would fail, "as it was too bad to blow people up in that way; it was not Christian." Whoever the culprit was, he cost the Confederacy the destruction of what was probably the most powerful warship afloat.[1]

A little-mentioned Confederate loss that can be attributed to this battle was the *Marion,* a side-wheel transport captured from the enemy. The evening prior to the battle she drifted from her anchorage at the mouth of the Ashley River and before anyone aboard realized what was happening her bottom scraped two of Rains's frame torpedoes. The explosions ripped her open, causing her to sink almost instantly—"in twenty seconds" according to General Rains. The *Marion* was the first major loss to mines for the Confederates. Somewhat later the transport *Etiwan* ran afoul of an explosive near the same place, but was run aground and saved.[2]

Because of their failure to reduce the Confederate defenses and their alleged lack of aggressiveness in the face of the line of torpedoes, the officers of the Union flotilla received a stinging rebuke in the press. C. C. Fulton of the Baltimore *American,* who observed the battle, sent a dispatch with the following comment: "The ghosts of rebel torpedoes have . . . paralyzed the efficiency of the fleet . . . and the sight of large beer barrels floating in the harbor . . . added terror to overwhelming fear. . . . The torpedo phantom has proved too powerful to be overcome. . . ." The Navy was furious with the statement, and Rear Admiral Samuel F. du Pont, who was in command of the South Atlantic

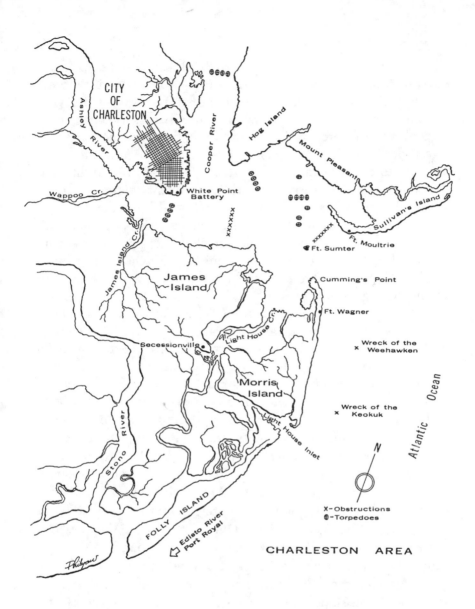

CITY
OF
CHARLESTON

Ashley River

Cooper River

Hog Island

Mount Pleasant

Sullivan's Island

White Point Battery

Wappoo Cr.

James Island Cr.

Ft. Moultrie

Ft. Sumter

James Island

Cumming's Point

Ft. Wagner

Wreck of the Weehawken

Secessionville

Light House Cr.

Morris Island

Wreck of the Keokuk

Light House Inlet

Stono River

Atlantic Ocean

N

FOLLY ISLAND

Edisto River
Port Royal

X – Obstructions
⊕ – Torpedoes

CHARLESTON   AREA

Blockading Squadron, voiced his resentment so shrilly that an inquiry to determine the truth of the charges resulted. Of course they were exaggerated, but the cold fact remained that torpedoes were becoming accepted as an important weapon of war.[3]

The Confederates in Charleston probably read the Baltimore paper before the Yankees who were still in the ships off the coast had access to it. The result was an immediate increase in their efforts. Captain Gray was sent to Skull Creek—one of those strikingly named waterways leading to Charleston—with a crew which had orders to float several mines. Laboriously ferrying supplies across swamp and marsh, the little band succeeded, after several weeks of preparation, in dropping the torpedoes on Tuesday night, May 18, 1863. The next evening they tried to entice the Yankees to run over the area in which the explosives lay hidden. Three small boats, carrying a total of twenty-eight men, approached Pope's Island which was garrisoned by Federal troops. Fifty yards from shore they were challenged; they replied with volleys of buckshot from double-barreled shotguns. The enemy retreated and the Confederates skirmished. To their disgust, the U.S. warships refused to take the bait and did not come to the aid of the beleaguered soldiers.

The vessels were tempted again on Thursday evening, but once more they remained stationary. The Southerners reported back to their headquarters with the recommendation that a cannon be sent over to open fire on the ships. "This," said the report, "would certainly bring a gunboat through Skull Creek of sufficient draft to explode our torpedoes. . . ." The enemy remained unperturbed.[4]

That summer Beauregard placed more confidence than ever in the ability of his torpedo units and sent them on numerous missions, most of which were routine. In July he instructed his engineers to set a quantity of torpedoes in Stono Inlet to the south and to put Rains's land mines in front of his lines at Fort (Battery) Wagner at the harbor entrance. Later, he saw to it that Light House Inlet, the Ashley River, Dill's Creek, and Wappoo Cut were supplied. Not all these projects were without incident. Not only was it necessary to move about in rowboats with the deadly explosives, but the crews were working in the very shadow of the guns on Union ships.[5] The Light House Inlet mission illustrates some of the difficulties.

Captain Gray made the "engines" and had them ready, but Colonel L. M. Hatch, who was in charge of the operation, was slow in making "arrangements for their disposal." He delayed for several days, using as his excuse that the tide was not right at night. On August 10, 1863, Beauregard replaced him with Major Stephen Elliott, a man who had been interested in torpedoes for a year and a half.[6] In fact, Elliott could be considered a pioneer in their development. Soon after the fall of Fort Pulaski, Georgia, in April, 1862, he had recruited several of his company (the "Beaufort Artillery") and set out some crude mines of his own design in the Savannah River. They were faulty, and nothing came of his effort. Since then, he had been following the progress of torpedo research and development closely.[7]

At 8 P.M. on August 10 (high water) the expedition left Secessionville below Charleston and proceeded down Light House Creek in three rowboats carrying eight torpedoes. After four hours of hard pulling they ended up lost in the myriad of creeks. Floundering in darkness all night, the crew returned to their base at dawn and rested while the officers detailed three additional hands. They set out again the next evening, and this time they reached their destination—a spot some four hundred yards from the Yankee fleet off Light House Inlet. Four lines of torpedoes (190 feet of rope with floats every ten feet and fifty-pound torpedoes on the ends) were set afloat by midnight. A close friend maintained that they were Elliott's invention; certainly their design was unlike those used earlier here, for they consisted of two tin cans fifteen inches in diameter, connected by a three-and-a-half-foot shaft. The lower of these cans contained sixty pounds of powder; the upper one was empty and provided buoyancy. Running down the connecting shaft was the barrel and firing mechanism of a sawed-off musket, loaded and cocked, and around the top of the float projected four spines which were tied to the trigger. The tripping of a spine released the trigger and the resulting shot exploded the magazine.

On their way back, Elliott and his men heard a sharp report. It was the detonation of one of the torpedoes which had drifted under the stern of the *Pawnee*, the same Union vessel that had scooped the first torpedo from the Potomac in 1861. A launch tied to her took the brunt of the explosion and was completely destroyed. Though shaken severely, the ship suffered only negligible effects. Her skipper dispatched his boats to scour the water for other floating mines, but they missed the one at the extremity of the line which had fouled on the *Pawnee*. It went off at 4 A.M., too far from the warship to hurt her, and was probably detonated by a piece of driftwood. A few minutes later the mortar schooner *C. P. Williams* spotted a pair of the cans floating toward her. Her cutter, directed by Ordinary Seaman John Jackson, tied a line to them and towed them off. At daylight Jackson hauled the "curiosities" ashore, and while engaged in this operation he heard two other torpedoes blow up in the creek. Undaunted, he removed the percussion caps from the shortened muskets and took the torpedoes to the *Williams* and the *Pawnee* where they were examined. Drawings were sent to Commander George B. Balch of the *Pawnee,* who hung the mines on his yardarms for the Confederates to see, at the same time ordering the recovery of still another that was seen floating nearby. The musket of this last mine had discharged and burst without exploding the magazine.[8] Balch had his sailors remove the powder from the lower container and added it to his stores "in part payment for his launch. . . ." [9]

A few nights later Elliott and his friend Lieutenant James A. Hamilton, C.S.A., went to set out four torpedoes near the *New Ironsides*. Elliott had selected Hamilton for the mission because "you understand the machines." They tried to reach the great ship several times, but were turned back by tides and heavy weather. The morning light a few days later revealed a raft of logs and nets around the *Ironsides,* an effective barrier which Hamilton erroneously thought was the result of spies' communicating their plans to the enemy.[10]

Rear Admiral John A. Dahlgren, who succeeded du Pont in command of the Union blockading squadron on the south Atlantic coast, became increasingly curious that summer about the obstruction across the channel between Forts Sumter and Moultrie. He made a point of interviewing every prisoner and deserter about it and summed up the results in a letter to Secretary of the Navy Gideon Welles in October:

There is another kind of impediment of which we know nothing with certainty—the Torpedoes—They may exist or they may not—may act or may not, when resorted to . . . may be anywhere. No one can give certain information in respect to them—some have seen such devices while being made—some think they are located in one place, and some in another—and there is on the whole, the greatest uncertainty in regard to them.

I have examined a number of deserters and prisoners—among them watermen accustomed to the harbor, and one Pilot—they all concur as to their ignorance of these Torpedoes.

All of these, however, will not prevent a suitable force from entering and penetrating to Charleston.[11]

What really bothered the admiral was not a fear of inability to cope with this menace but the fact that the Confederates frequently managed to gain knowledge of his attempts to learn more of the torpedoes. One incident so exasperated him that he published the following circular which was read to all hands:

The following has appeared in the Charleston *Mercury* of the 19th September: "A Yankee letter from off Charleston says:—Ensign Benjamin H. Porter; of the New Ironsides, who had been detailed for special service by Admiral Dahlgren, on account of the high character he has obtained in the Fleet, and his daring bravery and prudence, performed a feat on Monday night during the famous bombardment, that will ensure his high commendation by the Admiral. The duty assigned to him was to ascertain the nature and character of the obstructions across the harbor of Charleston, between Sumter and Moultrie. He had been up in one of the Ironsides' cutters, with a picked crew, for several nights on this mission, and was prevented from accomplishing it by encountering the picket boats of the enemy. On Monday night he was scouting around Sumter, when the furious cannonade commenced. All eyes were centered on the forts, and the work they were doing. But Ensign Porter saw the opportunity for his work had arrived. The flashing of the cannon from Sumter and Gregg guided his movements and he was enabled to reach the obstructions without being observed. He spent fully half an hour on them, thoroughly investigating their construction, and moved on toward the fleet in time to reach it at daylight. He immediately reported to the Admiral, who declared himself highly gratified at the information, declaring that he knew all that he desired to know. The information is, of course, kept secret, but will be availed of by the Admiral in a few days.

"From subsequent events, or rather lack of events, it is to be inferred that 'all he desired to know' was that his Monitors would be blown higher than a kite if he attempted to run in."

Here are three persons, one of them in our midst, who voluntarily bestow information on the enemy, of the highest importance, and which teaches him how to defeat us in reaching Charleston.

There is probably no means upon which the enemy here so relied for information as this insane propensity for making public the most valuable items.

I deeply regret that it is not in my power to treat this evil as it merits—it is unfortunately one of the besetting sins of the day—but

I call on every true man in this Squadron, to assist in bringing the perpetrators to . . . punishment.[12]

Two of the South's leading military inventors were called to Charleston that summer of 1863 to assist in the battle of the torpedoes. The first, Captain Ambrose McEvoy, C.S.A., had perfected fuses for artillery shells. A native of Ireland, he was raised in the South, and his first successful fuse was brought to the attention of Robert E. Lee early in the war. On examining it, Lee referred the inventor to Commander John M. Brooke. Subsequently the device was adopted into service "by the millions," and turned out to be one of the most successful inventions of the war. McEvoy also turned out several types of torpedoes. In August, 1863, one of his models was sent to Charleston, where it was demonstrated.[13]

The other inventor, Gabriel Rains, went to the Palmetto City on September 2, 1863, to "assume special charge of the preparation and location of torpedoes. . . ." He was to give special attention to Fort Wagner, the large and powerful earth fort which the Confederates had built across the narrow neck of Morris Island at the harbor's mouth. Wagner was important to the Union plan to encircle the harbor on its surrounding land areas, and strong pressure was being exerted against it. Its capture would put the Federals in control of much of the southern shore of Charleston Harbor.

This was a perfect spot to use land mines—it offered sand (which covered them easily and hid traces of the holes), unobstructed fields of fire from the fort, and no way for the Federals to flank the defenses since the ends of the lines rested on water.

The first shells equipped with the Rains fuse were laid on July 10, 1863, while Rains was in Mississippi. Some were exploded during an assault on July 18, and more were promptly put out. Fifty-six 10-inch shells were converted and buried on July 21, and others, forming a J-shaped defense around a marsh between the lines, were added later. Keg explosives were laid along the curve in the narrow neck between the marsh and the sea, and the shells were strung on the hook to Vincent's Creek, near the Confederate works.[14]

The Federals ran afoul of these torpedoes during the assault of August 26, 1863. The first casualty occurred when a corporal of the Third U.S. Colored Regiment stepped on one. The explosion tossed him twenty-five yards, stripping him of his clothing. He landed on another mine which failed to detonate. At first it was thought an artillery shell or hand grenade had struck him, but when the body was removed the second torpedo was found, giving rise to rumors that the Confederates had tied the Negro to it, so that the mine would explode when the corpse was moved.

Knowledge that these weapons were being employed there explained what had been called "one of the greatest mysteries in the defense of Fort Wagner"—why no chevaux-de-frise or other obstacles had been thrown up between the lines by the Southerners, and why they had retreated so obligingly from certain forward areas.

Eight more torpedoes were found on August 27, and their positions were carefully marked. Sappers cautiously removed them and rendered them harmless by boring holes through the casings and pouring water inside; sharpshooters vainly tried to explode others at long range the next few days. On the night of the thirty-first another was set off by a soldier crawling up to an advanced trench. It injured him and three other men.

More rumors swept the lines, describing torpedoes attached to penknives on the ground, designed to explode when the knives were picked up. Some of the floating mines broke loose from their moorings and drifted ashore within Union lines. One was discovered by an enlisted man who touched it "rather tenderly," trying to figure out just what it was. "I did not know . . . what the damned thing would do next," he explained candidly. Another infantryman got a saw and began to cut the wooden case open while his comrades watched from a safe distance. Suddenly, man, saw, and torpedo disappeared in a blinding flash.

Still another Yankee chanced to step on a bit of board. He was blown several feet in the air, losing a leg, a victim of the Confederates' favorite way of disguising mines. They would bury the barrel casings in the sand, fuse-side up; resting lightly on the fuse would be a board held in place by the sand.[15]

But the sappers had the greatest difficulties with the explosives, for these men, employed in digging trenches of the zigzag "sap" type toward the enemy's fortifications, ran into the mines more frequently than other soldiers. "A log troubles me in digging," said one. "Well, then, dig around it; don't bother me about it," replied his superior testily. The explosion blew the enlisted man "to atoms," for he had mistaken one of the beer barrel torpedoes for a log.[16]

Eventually the Federals crossed the minefield, and Fort Wagner was captured when the Confederates abandoned it under the cover of darkness. As at Yorktown, the Federals saw torpedoes everywhere, and hapless Confederate prisoners were forced to find them and dig them up. Casualties from these weapons during the battle were few, but their mental impact was great. One writer gave an apt description when he said that the torpedoes "attack both matter and mind." [17]

Though the torpedoes had no major effect on the outcome of the operation, the Confederates were pleased, for they believed that the mines appreciably delayed the enemy's advance. President Davis acknowledged "with gratification, the success of General Rains' subterra shells," and Beauregard was equally complimentary.[18]

It was a natural result of the battle that Rains should extend the scope of his invention. The network of roads converging on Charleston was a source of considerable worry to its defenders for several reasons. It was possible for small bands of enemy cavalry to slip through the pickets and conduct quick, sharp raids on positions, outposts, and buildings along the highways. There were more openings than there were units to watch them, but Rains had an answer. He sent small detachments consisting of only an officer and twenty-one men to implant these roads with mines. The explosives were placed unarmed in the highways and in that condition were, as Rains said, "perfectly harmless to citizens, until the enemy approach, when the shells can be primed in a moment for their reception. I am confident," he continued, "that if the enemy are once or twice blown up . . . raids ever thereafter will be prevented." [19]

In Virginia a similar plan was suggested. Congressman A. R. Boteler reported that General J. E. B. Stuart and the famous "Ranger," Captain John S. Mosby, had demanded some torpedoes in August. They intended to place the mines under the tracks of the Orange & Alexander Railroad, where they would be exploded by passing trains. The Secretary of War referred the plan to

Ordnance Chief Josiah Gorgas, who reported rather irritably that though he could furnish the weapons, he hoped they would reconsider, because derailments would "only irritate the enemy, and not intimidate him." [20]

At the same time, floating mines in Virginia streams were receiving a good bit of attention from the Federal Navy. Lieutenant Commander J. H. Gillis, skipper of the U.S.S. *Commodore Jones,* reported in mid-October of 1863 that for several days his ship had been dragging for torpedoes in the York River "near Rowan's [Roane's] Point, just above the mouth of Potopotank Creek" some seven miles below West Point. Though none were found, he was positive there were some in these waters; reports from a number of sources claimed that at least three dozen had been set out. This intelligence had been received in a number of ways: once, a group of men was seen at work in the river. At the same time, two Pamunkey Indian scouts in Gillis' employ passed through the country bringing out rumors to the effect that the Confederates contemplated mining the York. But try as he might the commander could find no explosives in the river. Later, a Southerner's letter, intercepted on the Mississippi, told of putting out ten torpedoes in the Pamunkey River and reported that on October 13 another fourteen had been ready. Federal information was correct, only the location was wrong; the Pamunkey empties into the York at West Point, Virginia.[21]

More torpedoes were seen floating in Hampton Roads and, on examination, bore evidence of having been recently set in the water. The discovery caused U.S. Navy Secretary Welles to order extra vigilance, and he admitted to "some uneasiness" concerning the safety of his warships.[22]

Things were somewhat the same in Charleston Harbor while the battle for Fort Wagner was going on. In late August and September of 1863 floating torpedoes were anchored in "large numbers" across all the channels Federal ships might use to reach the city. Mine barricades ran from Castle Pinckney in the middle of the harbor to Fort Ripley, and from Fort Sumter across the main channel to Fort Moultrie. They were in Hog Island Channel, Folly Island Channel, and in several other waterways. Occasionally, the Federals would scoop some up, and Lieutenant Commander George Bacon of the *Commodore McDonough* towed several ashore in Light House Inlet, down the Atlantic coast. To protect his ship he stretched nets across the channel to catch others. All the mines Bacon captured were Rains keg torpedoes, and one was sent by Major General Quincy A. Gillmore to the West Point Museum at the United States Military Academy for study. It is still there, in excellent condition.[23]

Near the end of 1863 another attempt was made to attack the Union ships off Charleston Harbor. Captain E. Pliny Bryan, who had hoped to sink the *New Ironsides,* was thwarted by a torpedo boat attack. In October he turned his efforts to the blockading squadron in general. Bryan intended to use a pair of 150-pound torpedoes having Rains fuses, connected by three hundred feet of rope and suspended from buoys to a depth of eight feet. They were to be put out when the tide was running in the direction of the ships, with the hope that they would be caught on the bows, as the mines in Maury's attack of 1861 had been.

With Bryan went Lieutenant John T. Elmore of the engineers and Captain John H. Mickler, Company E, Eleventh South Carolina Regiment. They were to leave on October 28 and obstruct Skull Creek, but a shortage of oars held them up until November 2. Even then, they started so late that dawn brightened before the boats got to their first destination. The next night eight small Rains kegs were set afloat in the ebb tide within 150 yards of Federal picket boats. This was tricky work, for it had to be done in silence, in complete darkness, and without detonating the torpedoes. The Confederates retired and waited. Their report of the operation claims that they heard at least one explosion that night, but Federal records do not mention any damage. Evidently, the blast was caused by driftwood detonating a torpedo.[24]

CHAPTER VII

CAPTAIN LEE'S TORPEDO RAM

Of all Southern commanders in responsible positions, Beauregard was perhaps the most receptive to trying new weapons when they seemed to have some basis of practicality. This is borne out by the fact that Charleston saw the combat experiments of more new implements of war than any other battle area of the Civil War. In fact, Beauregard even used one of his own ideas with excellent results on cannon in the harbor—a "rack and lever" mechanism by which heavy guns could be easily and rapidly traversed.[1]

Originator of many of the innovations used at Charleston was "an intelligent young Engineer officer" (Beauregard's description) serving on his staff, Captain Francis D. Lee. Born into one of Charleston's best families in 1827 and an architect by profession, Lee had assisted in the construction of Fort Wagner on Morris Island, but had lately interested himself in explosives and in experimentation with fuses. These were not of the ordinary powder-train class, but an advanced type: chemical fuses. They consisted of small lead tubes capped with hemispherical heads of very thin metal. Inside, he placed hermetically sealed glass phials containing sulphuric acid. Between this and the metal casing was a layer of a composition of his own invention: a combination of chlorate of potassa, powdered sugar, and fine rifle powder. Several of these fuses screwed into a container of powder (50–75 pounds) made a most dangerous weapon, for if the rounded cap were dented enough to crush the phial, the acid, acting on the composition, would ignite the powder and cause a sizable explosion. The strategy was to float the torpedoes across the channel where an invading Yankee ship might bang one against her sides—hardly an original idea.[2]

After establishing his weapons, Lee approached his commander with an idea for another unique destructive. The young engineer's new brainchild, buttressed with rolls of plans, was a strange-looking warship, one that would be more under the water than above. The idea came about while he was trying to figure out a way to take his torpedoes *to* enemy ships; if the Yankees would not come in, he would go out to their patrol stations.

The vessel was long, and low—so low, in fact, that water would wash across her armored turtleback deck. The crew and machinery were encased within the iron hull, mostly below the waterline. All that would appear above the surface was an expanse of curving deck, a smokestack, and a pilot's cockpit at the stern, also heavily armored. A long wooden spar, raised and lowered by a windlass, projected from the bow. The deadly torpedo, fairly bristling with fuses, was attached to its tip. The "torpedo ram," as Lee called his ship, would approach her target under high speed, lower the spar until the torpedo was several feet underwater, and ram it against the hull. He felt that the vessel would be far enough away to escape damage from the explosion and that her armor would deflect gunfire.

The general was very definitely interested, and Lee was soon on his way to Richmond, equipped with an order to take the plan to the Secretary of War. He returned empty-handed. The War Department and the engineers had brushed him off. This, they said, was a matter for the Navy, but that service lost interest on learning that Lee was an Army officer, and there the matter lay.[3]

Beauregard turned to another source. Certain states were build-ing warships, and he knew of some vessels for South Carolina that were being constructed in Charleston itself. Accordingly, on October 8, 1862, he wrote Governor Francis W. Pickens about Lee's ram. It really should be built, he said, for "I believe [it] would be worth several gunboats." [4] Giving Pickens but a short time to ponder this thought, Beauregard approached another person of power in the South Carolina administration. Four days later, he made a request of J. K. Sass, chairman of the state gun-boat committee. If we cannot build gunboats, he said, why not put the materials we now have into a ram which "can be ready sooner, will cost less and will be more efficacious"? [5]

Beauregard was a man who would leave no stone unturned to get what he wanted. On the following day Captain Lee was back on the train for Richmond with his plans, a scale model, and a letter from Beauregard to Adjutant General Samuel Cooper tucked in his bag. Because of the already deteriorating state of the Confederacy's railroads, Lee was forced to travel a circuitous route: via Columbia and Charlotte to Greensboro, North Carolina, changing for Raleigh, then from Weldon to Petersburg, where he took the "accomodation train" on the Richmond & Petersburg Railroad. Alighting at the terminal on the corner of Byrd and Eighth streets at 6:30 P.M. on October 16, 1862, he hired a carriage to the home of Congressman William Porcher Miles. The reason for the call was obvious: Beauregard wanted the support of the influential congressman. Miles was ill and in bed when Lee called, but he received the young man the next morning and scribbled an introduction to General Cooper.

The letter which Lee presented to the Adjutant General was a document couched in the most persuasive phrases that Beaure-gard and his aide, Alfred Roman, could devise. It was filled with rhetoric, logic, and patriotic homilies. The ram had the writer's "hearty approbation, as offering altogether the most practicable means of a successful encounter with the formidable ironclad gunboats of the enemy I have yet seen." The state government had been so impressed it had allocated $50,000 to the ram's construction. "Practical builders" had approved it, but Beaure-gard was sending Lee to Richmond to get the aid of the central government for material. In addition, he hoped other such ships would be constructed upon orders from Richmond. They would cost a third less, would be built in a third of the time required for gunboats, and would be fully as destructive as the larger vessels. If materials in Charleston could be applied to the ram, Beauregard felt certain it would "render signal aid in holding this fort for the Confederate States." [6]

After studying the letter, Cooper stared at the young captain, his brow furrowed. Upon reaching a decision he signaled a clerk, but directed his remarks to the engineer. The idea impressed him, he said, and he was going to submit it to Secretary of War Randolph for consideration. (Lee reported Cooper's reaction as one of "warm interest.")

It was a simple matter to follow the general's aide across the hall into the office of the Secretary, but getting the officials to act was much more difficult. Instead of exhibiting interest, the Secretary passed the letter over to Colonel J. F. Gilmer of the engineers without action. Lee was told to return the next morning.

When the doors of the War Department were opened for business Saturday, October 18, the first callers were Captain Lee and Congressman Miles, who had devised a plan to cut the project from the confines of red tape. Miles rescued the papers from Gilmer and personally saw the Secretary. This time Randolph read them with care, decided this was a matter for the Navy, and drafted a warm recommendation. Everyone liked the scheme, but no one was willing to take the initiative.

Lee and Miles decided to take it to Secretary of the Navy Mallory themselves. Mallory's office was but a few steps—upstairs over General Cooper in fact. He made a "careful examination of the design, expressed his deep interest" and to the joy of the petitioners "his entire willingness to furnish everything in his power to make its accomplishment as early as possible." Mallory called for Commander John M. Brooke, Navy Chief of Ordnance, and John L. Porter, the department's naval constructor—the men who had devised and supervised the building of the C.S.S. *Virginia*. Lee and Miles may well have felt a bit nervous during the following hours, for the naval designers certainly were aware of Beauregard's repeated criticisms of their ironclads and his hopes of replacing them with this ship. After a thorough examination of the plans, they gave a favorable, if not specific, recommendation, going on record as "approving . . . the design as offering a valuable auxiliary to the defense of rivers and harbors."

Elated, Captain Lee and Congressman Miles rushed back down to the War Department and closeted themselves with Colonel Gilmer, whose matter-of-fact attitude of Friday was replaced by sudden and deep interest. Gilmer pored over the charts, gave "valuable advice in reference to . . . details" and advised Lee to institute further experiments. He was careful to point out that work could start at once.

It had been a day full of achievement and the pair were satisfied. Only one more step was needed to complete the program: an order bearing Secretary Mallory's signature for the release of armor plating, machinery, and other materials needed to build the ship. Instead of waiting until Monday, Lee left this matter in Miles's hands and boarded the train that same evening. He would regret his haste in future months.[7]

Back in Charleston the wheels began to turn. In paragraph three, Special Orders No. 210 of October 31, 1862, Headquarters, Department of South Carolina and Georgia, Beauregard made the project official, ordering Captain Lee to take charge of construction with full power to "examine and supervise all accounts" and to spend the $50,000. F. M. Jones, a ship carpenter, and Cameron & Company, Charleston machinists, were to do the work. Jones had been suggested by naval constructor Porter, who was familiar with his work. According to Lee's understanding, the Navy was going to make engines and boilers available for the ram.

When word of this project spread among the services, several young and enthusiastic "torpedo minded" officers scrambled to join, pressing for the establishment of a special corps to be recruited from the various branches to sail the ship and others like her, and requesting that Captain Lee head the organization. Among the men were Isaac Brown, of Yazoo River fame, and Lieutenant William T. Glassel, C.S.N.[8]

Work went well the first week of construction. Jones and his men swarmed over an uncompleted gunboat frame, hammering and sawing, "filling in" the bare ribs and changing the shape of the bow and stern to suit the new design. Piles of Carolina pine cut to size for the deck timbers lay at hand, though Lee entertained the hope that he could get oak from the Navy.[9]

With his request for oak (18,500 feet of 4-inch planks) Lee submitted other material requirements to Captain Duncan N. Ingraham, who represented the Navy in Charleston. These requests included 6,000 pounds of oakum, 5,000 pounds of 8-inch spikes (½-inch size), 5,100 pounds of 10-inch spikes, 10,500 pounds of ¾-inch iron plating, 4 tons of coal, and five good ship carpenters. Besides this, Cameron & Company needed a great deal of iron: 60 tons of cast iron bars, 10 tons of 1½-inch bars (for bolts), and various small quantities of sheet iron and copper. This was probably material for the hull, which had been requested with the idea of building a stockpile nearby upon which the construction crew could draw.[10]

Ingraham took so little time to consider the application that his reply was on Lee's desk that same afternoon. The naval officer was sorry, but the Navy had absolutely nothing in its yard except a few three-quarter-inch iron plates and a handful of spikes, nor was there much chance of getting the supplies elsewhere. There simply was not enough of these sorely needed items to go around. (Lee was already regretting not having Secretary Mallory's signature on an order requiring naval officials to supply material on request.) Captain Lee decided the next move would be an attempt to get the iron from either the Atlanta or the Etowah Iron Works, using some of the original $50,000.

Beauregard was impatient. Here, in hand, were the plans for the ship "destined . . . to change the system of naval warfare," but its very construction was being held up by disinterest.

As Thanksgiving approached, Lee worked his men, including house carpenters, with great fervor. The new bow frame was put in place while torpedo and windlass designs were laid out and awaited iron for fabrication. The woodwork was going nicely, but where was the machinery the Navy had promised? Letters brought only vague replies, and Lee sent a man to Richmond who secured the release of an engine, boiler, propeller, and shaft, but he needed another set. (Lee preferred built boilers because it took so long to have them made to order.) As a reserve, there were some boilers in Wilmington, North Carolina, designed for a ship of similar size which had remained unbuilt because of an epidemic of malaria.[11]

The iron did not materialize, and Lee fell back on an emergency plan and scoured plantations along the Cooper River for scrap. A total of twenty-five tons—parts of old cotton gins, buggy axles, porch railings, skillets—was requisitioned. Some "much worn" machinery that had been in the old tug *Barton* at Savannah was purchased for the ram.

Then came an unforeseen problem—a strike. The Confederacy did not recognize work stoppages as legal, and short shrift was usually made of them. The spiral of inflation gripped the South; its currency dropped in value every day. Occasionally workers left their jobs in attempts to get raises so that they could keep up with the cost of living. Such was the situation at Charleston when, without warning, all the men working on Lee's ram dropped their tools one morning and walked off. Captain Lee rushed out and, with the hull of the ship above him, stood on a pile of timber, pleading with the men to return. When they refused he told them the probable consequences: they would be drafted into the Army where their services would bring only a fraction of what they were being paid for building the ram; they would have no choice of what kind of work they would do. The workers talked among themselves and, one by one, climbed back on the ways. A small knot was left—the free Negroes. Lee had no threats for these men, for they were not subject to the draft. One stalked off, but the rest joined their white comrades on the ship.[12]

By New Year's Day the ram's lines were beginning to take form. The hull was nearly finished, her iron prow was being cast, and friction tubes for the torpedoes were being made at Charleston Arsenal. A boiler had been installed and the *Barton*'s machinery was on the ground, waiting its turn.

But when it seemed that success and completion were near at hand, work slowed. Cameron & Company was taken over by the Navy and the old, easy, direct coordination stopped. All matters relating to the ship had to go through channels and took twice as long. The Navy still had not sent any armor plating and, it seemed to Lee, was trying to delay the project. Beauregard's inspector general, Lieutenant Colonel Alfred Roman, visiting the scene on March 9, 1863, was somewhat disappointed with what he saw. Lee was working sixty-one carpenters a ten-hour day, but a great deal still remained to be done. In fact, Roman feared that the ram—if finished at all—would be finished too late.[13]

To add to his apprehensions, word of a Union naval attack, received by reading enemy signals with a captured key, was relayed to the city. Among the alerts was one sent to Captain Lee, ordering him to make "all the necessary arrangements . . . to insure the complete destruction of the torpedo ram . . . at a moment's warning." The attack was repulsed, but work on the new vessel slowed even more. By mid-April Roman's fears were realized, and all construction came to a halt. The work had gone as far as it could without the installation of the long-promised plating. Though the Navy would send none for the ram, it had supplied enough so that the ironclads *Chicora* and *Palmetto State* were nearing completion. Beauregard angrily wrote General Cooper and told him that this project was not something that he alone favored; even Captain John R. Tucker, C.S.N. (who had succeeded Ingraham), was for it, declaring "unhesitatingly that this one machine of war . . . would be more effective . . . than nearly all the iron-clads here, afloat and building. . . ."

"I do not desire to impose my views," continued Beauregard untruthfully, "but feel it my duty to urge an immediate investigation by a mixed board of competent officers, to determine whether it be best for the ends in view to continue to appropriate all the material and employ all the mechanical labor of the country in the construction of vessels that are forced to play so unimportant and passive a part . . . in the future as in the past." Even this plea was not effective. Dejectedly, he noted: "I have written and telegraphed . . . until my hand is hoarse." [14]

One of the reasons for Captain Lee's inability to devote his full energies to his ram during that winter of 1863 lay in a concurrent scheme contemplating the utilization of torpedoes for a single swift blow against the enemy. First estimates of the time for building the ship had been three months. When the deadline approached in January, the ram remained far from finished. Still, some people thought the idea was sound and the spar torpedoes were probably effective weapons, but how could they be put into action quickly? The ram project was going to take a long time.

One possibility was to arm existing warships with torpedoes. The spar mechanism was installed on the bow of the clumsy, slow, ironclad *Palmetto State*. Her boom was twenty feet long, and the windlass was in the armored casemate. The torpedoes themselves were kept ready for instant use; every two weeks one was re-moved and examined—a ticklish job as verified by Captain W. H. Parker: "As executive officer I always attended to this with the gunner, and it was no joke to do it. In the first place we had to go out in a boat and take the torpedo off the staff, and in rough weather it was hard to keep the boat from striking it. . . . A moderate blow was sufficient to break the glass phial inside the fuses and cause an explosion. . . ." [15]

This was but a hastily conceived, temporary measure that pleased no one. What was needed was an offensive against the enemy fleet that could bring important results. Much of the impetus for such an attack came from very high sources indeed. Secretary Mallory himself, on February 19, 1863, wrote to Lieu-tenant William A. Webb, C.S.N., a series of suggestions for an expedition of sailors in small boats sent to board Union monitors and douse them with incendiaries. Webb went to Charleston and set about organizing his forces. When Lee and Glassel heard of the project, they recommended that the spar torpedo be used instead. Lee immediately requested that Beauregard let him try some experiments in the harbor to see if this idea would work or if the frail rowboats would be destroyed in the explosion. Lee thought not. "I believe that the entire force will be expended through the sides of the vessel [attacked]," he felt, "for the reason that this is the only compressible substance in contact with the torpedo, the water surrounding it being perfectly non-compressible and not yielding except by actual displacement, which requires a certain lapse of time to overcome inertia of rest. . . ." [16]

An abandoned ship was located and turned over to him. His calculations were based on the torpedo's exploding more than six feet below the waterline. Since the hulk's draft was insuffi-cient, she was weighted with rubble from Charleston's "burnt district"—an area that had been gutted by fire. Filled and resting in the water up to her gunwales, the ship drew only six and a half feet at her bow. Lee had hoped for a foot more, but this would have to do. His torpedo boat was a flimsy "light-built canoe" about twenty feet long, having a twenty-two-foot spar tipped with a thirty-pound torpedo. On March 12, 1863, he rowed into position and waited for high water which was expected at 1:30 P.M. A northwest breeze freshened into a small gale and the water roughened; the boat began pitching violently. As she rose and fell with the swells and the torpedo slapped the water, Lee feared that the impact might break his seal and wet the powder. His crewmen had greater fears—they were afraid the thing would explode.

At 2:30 P.M. a ship's cutter was called which took them in tow while Lee payed out line attached to the hulk. In position again he secured the rope to the bow of his boat, tied another to her stern and gave the free end to the cutter. By straining the cable the torpedo boat stayed on course, and the torpedo apparently struck its target. There was no explosion. Finding the torpedo floating against the keel, Lee pulled it out gingerly and put it in his boat, calling off further experiments that day.

Lee tried again at 8 A.M. on March 13. This time he was rewarded with complete success. The torpedo produced a muffled roar, the target vessel trembled, and a geyser of water, dark with mud, rose beside it. When the water subsided, the ship settled and disappeared in twenty seconds. The torpedo boat was undamaged.[17]

Among the officers watching the experiment was Captain Glassel, who was deck officer of the ironclad *Chicora*. Glassel was greatly impressed and decided that a fleet of these boats would be able to deal a crippling blow. He went to his superior, Ingraham, and said that if forty of these boats were sent out, he would like to lead them into action. The only additions they needed were iron shields to protect the helmsmen from rifle fire. Ingraham refused. Glassel's "rank and age did not entitle [him] to command more than one boat," regardless of size.[18]

Ingraham was under double assault over the matter, for while Glassel was talking, Beauregard was writing, asking that several boats be fitted out, "not only to defend the harbor against the ironclads but to blow up the blockaders at night." The result of the discussions was that Glassel was allowed to attack with a single torpedo boat. Glassel was somewhat pleased, but Lee was vexed for he disapproved of sending a lone boat. The scheme was based on surprise and the need for simultaneous attacks on several ships. A single assault would do little more than alert the enemy, he felt. His protest fell on deaf ears; higher authority maintained that one boat had a better chance of slipping through the guard line.[19]

One night Glassel and a crew of six volunteers took Lee's "canoe" into the outer harbor toward the 2,100-ton, paddle-wheel blockader *Powhatan,* whose lights were clearly discernible from Fort Sumter. They timed their venture with the ebb tide at 10 P.M. and rowed quietly. The moon had set and the sea was quite smooth. When they were some six hundred feet from the *Powhatan* a voice sounded the usual hail: "What ship is that?" Glassel replied with "evasive and stupid answers" hoping to confuse the Federal lookout until he was close enough to strike. The ship threatened to fire on him unless he stopped and identified himself. At this point one of the oarsmen "from terror or treason" backed his oar and stopped the boat, causing it to drift past the warship. Realizing that the enemy was now alerted, Glassel prudently cut his torpedo adrift and pulled away. The recalcitrant oarsman later deserted.

Because of the lack of success during this maneuver, a number of officers who admitted to having "torpedo on the brain" became dejected over future possibilities. Before any new moves could be undertaken, Glassel was transferred to Wilmington, North Carolina, to supervise the equipping of the new ram *North Carolina,* and Ingraham was replaced by Tucker.[20] These changes did not have a really serious effect on plans for using the spar torpedoes, however. Secretary Mallory pressed demands for the training and organization of Lieutenant Webb's force which had become known as the Special Service Detachment and was being commanded by Lieutenant W. G. Dozier. Sailors from Charleston and Wilmington were recruited, and ten boats were procured. The small side-wheeler *Sumter* was assigned to them as their command ship. Webb directed the operations from Richmond and cautioned Dozier to "be careful to select the coolest and best men under your command to discharge the torpedoes. . . ."

Rigorous training soon had its effect, and the men lost their fear of and, according to Captain Parker, their respect for the new weapons: "It was not at all uncommon to see a sailor rolling down to his boat, when they were called for exercise, with a quid of tobacco in his cheek and a torpedo slung over his back; and when it is recalled that each . . . had seven sensitive fuses which a tap with a stick or blow with a stone was sufficient to explode and blow half the street down, it can readily be believed that we gave him a wide berth." [21]

Their plans came to completion on April 9, 1863, two days after the Federal ironclads' great attack on the shore batteries. Webb came down from Richmond and held a conference with Beauregard and Tucker to discuss how they might deal with the enemy ships that now lay nursing battle wounds just outside the harbor.

This was the beginning of the first planned mass attack with torpedo boats. Parker was detached from the *Palmetto State* the next day on orders to attack the monitors with the whole squadron of spar torpedo boats. The outline was as follows: the boats would rendezvous on the first calm night at the entrance of the creek just behind Cumming's Point at the harbor entrance; they would coast past the point to a spot near the Union ships (six monitors and the *New Ironsides*); there, they would form a line, lower the torpedoes by windlass, and attack in pairs. If hailed they were to reply an ambiguous "boats on secret operations" or "contrabands." Beauregard was sanguine concerning their chances of success. He wired Senators James L. Orr and Robert W. Barnwell that the expedition would "shake Abolitiondom to [its] foundation if successful." [22]

Fifteen boats (five had been added to the group during the planning stage) assembled around the steamer *Stono* in the darkness of April 12, 1863. Parker passed orders to the officers that they would attack at midnight. Just as they were about to leave, Commodore Tucker came aboard to report that the Federals had vanished—not a ship was in sight! There was no other course but to cancel the undertaking. Years later Parker revealed that he, personally, was glad that the attack had been called off because of the "many apprehensions" he had secretly entertained about the frail and leaky craft they were using.

The next day new plans were proposed by Beauregard. He suggested that four or five small steamers, burning smokeless coal and towing four torpedo boats each, go to the harbor entrance, cast the boats off, and let them attack the blockade fleet. Before this idea could be seriously considered, a telegram from the Navy Department ordered the men back to their ships. Beauregard tried to intervene, but to no avail. Lieutenant Webb went back to

Richmond, leaving Dozier with eleven officers and forty men. Still the Creole general tried to get the attack under way, hoping that "not one of the monitors will ever get away again." [23]

Some of the enemy ironclads had taken refuge in the North Edisto River down the Atlantic coast, and six of the little boats were sent out on May 10 to scout them. Parker had the boats towed by a tug through Wappoo Creek from the harbor into the Stono. There they cast off and rowed the rest of the way. General Johnson Hagood, who was in charge of this area, furnished intelligence as to the location of the enemy, and the boats were hidden in Pohicket Creek until the time was ripe to make an attack. On the twelfth, while they were making ready, a nose count of the crewmen came up one short. Since a quick investigation disclosed that one man had deserted, Captain Parker called off the venture and his boats were returned to Charleston by wagon. [24]

The boats saw only occasional patrol duty during the following months, and their crews were shipped away one by one. In early July the remaining men tried to preserve their status and addressed a petition to Dozier asking to stay with the boats. It had little or no effect. Once or twice they were ordered to stand by, but by September all of them were gone and the entire project was abandoned. [25]

It was to the south, however, that a large ship first tried to smash a spar torpedo against a Federal warship. The move was directed by Webb, and the ship he used was the ironclad ram *Atlanta*, the strongest and most powerful ship in the Confederate Navy. Originally built as the iron steamer *Fingal* in England, she had taken refuge at Savannah in November, 1862, after the Union capture of Port Royal, South Carolina. Her upper works were cut away to within two feet of the waterline, and an iron-sheathed turtleback was built on the hull. This structure was covered with wrought iron plates four inches thick and was pierced for eight guns, but only four—two 7-inch and two 6.4-inch Brooke rifles—were aboard. When finished and rechristened the *Atlanta*, she was 204 feet long and 41 feet wide. Her speed was said to be ten knots, a very fine achievement for an armored ship of that day.

The United States kept several ships in Wassaw Sound to nab

blockade runners going to and from Savannah, but when Admiral du Pont heard that the *Atlanta* and other ironclads might try to raise the blockade, he sent a pair of monitors, the *Weehawken* commanded by Captain John Rodgers and the *Nahant,* as reinforcements.

With an ample amount of chilled, armor-piercing shot and shell in her magazine, one torpedo on a spar at her iron beak and five more in reserve, the *Atlanta* made her bid and moved into the sound at 6 P.M. on June 15, 1863. She coaled all night and, with two wooden consorts, moved within five or six miles of the Union ships. At 3:30 A.M. on the seventeenth, Webb signaled his ship to get under way and drove at full speed for the nearest target, the *Weehawken*. At a distance of a mile or so, he ordered the torpedo spar lowered. His plan was to sink one Yankee with the torpedo and the other with gunfire; it was generally felt by both sides that the *Atlanta* was fully capable of accomplishing this feat. Suddenly, only three quarters of a mile from the target, she grounded. By reversing the engine, Webb freed her, but the ram refused to answer the helm and soon ran aground again. Pinioned, she presented a perfect target for the monitors and surrendered after the *Weehawken* fired half a dozen shells down her length.[26]

Lieutenant Commander David B. Harmony of the *Nahant* boarded the *Atlanta* with a prize crew. According to a Northern newspaperman, he changed flags and went to drop anchor; Webb stopped him: "For God's sake, don't cast off these anchors; we have a torpedo underneath the bow!"

"I don't care anything about your torpedoes," replied Harmony, "I can stand them if you can, and if you don't wish to be blown up with me, you had better tell me how to raise the torpedo!" Webb gathered those of his crew responsible for this operation and began working the windlass. "Soon there appeared coming out of the water a huge torpedo." Harmony directed his men to remove the caps carefully and pour water in them.

The Yankee correspondent commented on the *Atlanta*'s torpedo arrangement:

It is evident that the rebels have taught us a good lesson on the torpedo subject, as connected with ironclads . . . how a torpedo could be safely carried in front of a vessel without interfering with its steering and other movements, and be at the same time secure from any explosion until the proper time. The Atlanta's torpedo gearing solves the question. The forward part of the ram . . . is solid iron, twenty feet in length, and so overlaid by steel bars, with their ends protruding below . . . that a huge steel saw is formed. . . .

From the deck . . . just ahead of its juncture with the vessel, arises a strong iron bar with a pivot at its top, to which is attached a massive iron boom which runs just over the ram's prow, and then forming an elbow, it descends three feet below the water line, where it forms another elbow, and . . . some two feet . . . a . . . socket. . . . In this . . . is . . . another iron boom . . . and at its end . . . a torpedo, all capped and ready. . . . You can hardly conceive of a more perfect or effective engine of destruction. . . .[27]

Wilmington, too, soon had its torpedo boat. When Glassel was transferred from Charleston, he had carried his enthusiasm for the new weapons with him, and it was not long before he had outfitted a swift little steamer with the pole and gear so that he could attack the U.S.S. *Minnesota*—the most powerful blockader off Cape Fear. Just as he was about to make his move, he was sent back to Charleston.[28] In the Palmetto City he found ample evidence of the spar torpedoes' influence on Confederate naval plans, for the devices had been attached to all the warships in the harbor. Even more interesting was the fact that the work on Lee's ram had been resumed.

The great armored enemy ship *New Ironsides* had become the symbol of Union naval might in the eyes of the Confederates at Charleston, and plans for destroying her were many. After the failure of the small-boat attacks, Lee's ram, still lying uncompleted on her stocks, was reconsidered. This time, it was the Navy that tried to finish her. A board of naval engineers surveyed her, "to ascertain whether she was fitted for the purpose for which she was proposed." They reported favorably, noting especially that she should be able to turn up to six knots—twice the speed of the ironclads. Commodore Tucker became so interested that he looked at her himself and ordered the work begun in order to finish her as soon as possible.

Captain James Carlin, C.S.N., even offered to buy her; he wanted to attack the Federals and collect prize money. His offer was refused, but he was promised her command. Construction was rushed; a cutwater was put over the bow, and three torpedoes were placed on the spar (in the manner of the three balls designating pawnshops) in hopes that at least one of them would detonate. The ram was launched without fanfare, and a short shakedown cruise around the harbor on August 1 was a complete success. Satisfied, the Navy made her ready for combat.[29]

The goal for which Lee had argued and worked was now at hand. So impressive were his plans that the rich and powerful John Frazer and Company (which did a great deal of business with the Confederate government in shipbuilding abroad) formally offered to let him supervise the construction of another ram in England, which was to be paid for by the city of Charleston. Lee was anxious to accept, for overseas he would be able to build better ships in finer yards, with good materials and skilled workmen. On June 3, 1863, he formally submitted the plan to the engineer department, asking "to be detached for special service in England or elsewhere." In a telegram he attached the information that it would take "not more than six months" for the project and, in addition, reminded the department that he was the engineers' "Senior Captain" and, he thought, was due for his majority.

The file was referred to the War Department, but before a decision could be made, another telegram arrived from Lee: "Gen'l Beauregard instructs me to telegraph that facilities now offer for me to go without delay with naval officers. Will send my accounts [expenses in building the ram] by first opportunity."

This plan was disapproved in Richmond. Apparently it was felt that Lee would be more valuable where he was, supervising the completion and operation of his ram. Nor was he awarded a promotion; instead, he was ordered to concentrate all his energies on the ship, which he certified completed on August 3, reporting proudly that she was a good ship and "exceeded both in speed and seaworthiness."[30]

Lee's torpedo ram cast off her moorings just after dark on August 20, 1863. Nearly invisible, settling deep in the water with only her gray works above the surface, she glided toward Fort Sumter. As an added precaution, she was burning anthracite coal, which gave off a minimum of smoke and sparks. The fort was touched at ten o'clock, and a guard of ten soldiers, armed with muskets and commanded by Lieutenant Eldred S. Flicking, boarded the ram. The night was moonless; a gentle southwest wind raised small swells. Carlin passed the line of obstructions across the channel at 11:30 P.M. and discerned the lights of the *New Ironsides* and five monitors anchored off Morris Island. At a quarter-mile range he lowered the spar and began his attack course, keeping the target over the port bow. When the ram was about fifty yards away, Ensign Benjamin H. Porter, the *Ironsides'* officer of the deck, made out a strange vessel and gave the hail: "Ship ahoy!"

The Rebel soldiers took cover, pointing their rifles up at the deck towering above them. Carlin thought swiftly. So new was this mode of warfare that he risked giving his enemy a brief warning, for he was certain the ram would strike home. He replied simply, "Hello!"

"What ship is that?" shouted Porter. "The steamer *Live Yankee*," answered the Confederate. At that instant, some forty yards from the target, Carlin stopped his engine and ordered the steersman to put the helm "hard a-starboard"; the ram was approaching on an oblique course and would miss the *Ironsides*. The helmsman, whose position was below, did not understand and was slow to respond. Carlin excitedly repeated the command three times, but the tide took the ram in its grip and swept her alongside the ironclad.

From the deck of the *Ironsides* Ensign Porter, peering over the side, commanded Carlin to tell him from what port the intruder hailed. "Port Royal," was the answer. Sailors' heads began poking through the gunports of the Union ship, leading the Confederates to believe that they were being boarded. They prepared to defend themselves while Carlin dashed below to start the engine. At this critical time, it had "caught upon the center" and refused to budge.

Once again he was hailed, "What ship is that?" "The United States Steamer *Yankee*," yelled the skipper of the little ship, as drums aboard the big warship beat to quarters. A rocket sizzled from the *Ironsides'* deck warning other vessels, and marines lined the rails with rifles trained on the Confederates. On board the ram her engineers finally got the balky engine started, and she drew away while Porter demanded her surrender. To gain a precious moment, Carlin answered, "Aye, Aye, Sir, I'll come on board." The marines and small cannon were already firing in the general direction of the ram, missing her, as the *Ironsides* slipped from her anchor and commenced to back. The Southerners deduced that they had passed from view and, for the moment, were safe.

Captain Carlin momentarily considered attacking one of the monitors, but because of the recalcitrant engine and noise from the monitors indicating that they too were alerted, he decided to play safe and return to Charleston.

Chagrined over the failure of the engine, he dashed off a bristling report to Beauregard: "I . . . most unhesitatingly . . . express my condemnation of the vessel and engine . . . and as soon as she can be docked and the leak stopped would advise making a transport of her." Beauregard hated to lay up the ship for which he had made so many claims. "I feel convinced," he wrote, "that another trial under favorable circumstances will surely meet with success, notwithstanding the known defects of the vessel." He talked Carlin into staying on, but agreed that the ram should undergo improvement before she saw service again.[31]

The Confederates' secret was out. Within two weeks of the attack, Carlin's brash reply to Porter's hail was quoted delightedly throughout Dixie. Clerk J. B. Jones of the War Department wrote of it in his diary, exulting over the statement, confidently expecting greater things. (Actually, there was a *Yankee*. She was a 328-ton paddle-wheel gunboat mounting three cannon.) Admiral Dahlgren tried to learn why a strange ship was allowed to come up next to the most powerful vessel in his fleet with seeming impunity. Later, a Mr. Haynes of New Providence, in the Bahamas, told a Union officer just which boat it was, and, by the middle of September, Dahlgren had a fair description of her.

This was the ram's first and last combat service. She was never taken into action again.[32]

CHAPTER VIII

DAVIDS

Glassel had been back in Charleston only a short time when he learned of an interesting private venture, one that appeared to offer infinitely more possibilities of a successful coup against the enemy than the little row-powered torpedo boats. The Southern Torpedo Company, a group of Charlestonians headed by Theodore Stoney and Dr. St. Julien Ravenal, had built, with the assistance of Captain Lee, a small propeller ironclad specifically designed to deliver the spar torpedo against Federal warships. The strange little vessel was hardly more than a semisubmerged shell around a powerful engine. Fifty feet long, she was ingeniously constructed so that she could admit water into ballast tanks to lower her until her curved deck was awash. An armored funnel forward was her tallest feature, and, as a whole, she presented an extremely low silhouette. The torpedo was carried upon a 14-foot, hollow iron folding shaft at the bow. The weapons themselves were of an improved type, weighing about a hundred pounds, and were equipped with four supersensitive primers.

When he learned of this tiny vessel called the *David* (probably because of her diminutive size as compared to her prospective targets), Glassel promptly volunteered to command her. On September 18, 1863, his offer was accepted. The crewmen included one of the owners, Captain Stoney, who was first officer. Also aboard were Captain James H. Tomb, a resident of Savannah who had been in command of one of the rowboats, and Charles Scemps and James Able, firemen.

Their instructions were intentionally vague. The initial orders simply released them so they could report to General R. S. Ripley "for special service against the fleet of the United States off the Harbour," but on September 22 Glassel was specifically directed to "assume Command of the Torpedo Steamer 'David,' and when ready will proceed to operate against the Enemy's Fleet . . . with the view of destroying as many of the enemy's Vessels as possible. . . ." [1]

On taking charge of the boat, her crew painted her deck blockade-runner gray and took her to various parts of the harbor in test cruises. Several changes in the original arrangement occurred: the firemen were replaced by James Sullivan and Pilot Walker Cannon of the *Palmetto State;* Captain Stoney was assigned to build more "Davids"; and the ship itself was absorbed into the Navy.

More training followed before Captain Glassel felt his crew was ready. After careful deliberation, he chose the challenging might of the *New Ironsides* "to pay her the . . . highest compliment." (Its presence continued to rankle Confederate military leaders.) Perhaps it was not only her size but also the *Ironsides'* great armament that attracted the Southerners. She "threw a great deal more metal at each broadside than all of the monitors together," explained one of them. [2]

Conditions were ideal for the attack on the night of October 5, 1863. In a light wind and a faint mist the *David* left Atlantic Wharf at dusk on the ebb tide. Glassel consulted his watch from time to time and, as they neared Sumter, rang up four knots, hoping to time his attack with the change of tide at 9 P.M. Off Morris Island he saw the pinpoints of enemy campfires and could hear the murmur of soldiers' voices. The *David* did better than had been expected; at about 8:30 the *Ironsides* came into view.

Glassel stopped his engine and drifted. Then he heard the reverberation of the nine o'clock gun from the flagship, and one by one from the other ships came the piercing shriek of fifes and the rattle of drums as the Yankees secured for the night.

"It is now nine o'clock. Shall we strike her?" Glassel reportedly asked his crew. Cannon replied laconically, "That's what we came for. I am ready." Tomb was ready too: "Let's go at her then, and do our best!" Sullivan agreed: "I'm with you all, and waiting. Go ahead!" Grimly their Captain ordered full speed and attended to numerous duties. Double-barreled shotguns—one per man—were loaded with buckshot, their .36 caliber Colt "Navy" revolvers were checked, and the torpedo spar was lowered. Pulsating with the full power of her engine, the *David* boiled across the sea toward the unsuspecting Yankee.[3]

On board the great ship, Ensign Charles W. Howard, the officer of the deck, was pacing, carefully watching the sea. Since that strange incident of August 20, all ships were on the alert for a possible boarding expedition. At 9:15 a sentinel spotted the wake of a fast-moving vessel headed toward their starboard side. He alerted Howard, who leaped to the gangway, yelling, "Boat ahoy! Boat ahoy!" There was no reply. Howard could see no ship, only the phosphorescence of a wake. "What boat is that?"

At this instant the *David* came within range of small-arms fire, and Glassel raised his shotgun and pulled the trigger. Buckshot spewed through the air, felling Howard. At the same time, the torpedo smashed into the iron hull and exploded with a deafening roar. The huge ship trembled in every plate and spar throughout her length; rigging crashed to the deck; cannon leaped from their moorings; shot and shell scattered across the decks, rolling as the vessel heeled violently to port.

An enormous jet of water erupted from the sea and came splashing down on both vessels. Great waves swept over the *David*, drenching her crew and putting out the engine fires. As she settled deeper in the water, Glassel ordered his men to abandon ship. They stripped off their coats and dove over the side, wearing their cork life jackets. Glassel swam away while Sullivan took refuge in the enemy's rudder chains. Tomb, an accomplished swimmer, struck out for Morris Island; Cannon, an old salt who had never bothered to learn to swim, grasped a rope from the *David,* which was still afloat. Tomb reversed his course and climbed back aboard pulling Cannon after him. Out of range, they set about relighting the fires. Soon they were chugging by the ruins of Fort Sumter.

An hour later a picket boat from a Federal coal transport rescued Glassel and took him aboard the U.S.S. *Ottawa,* shivering and nearly naked. He was quizzed and gave only a minimum of information about the attack, but some papers found on him provided the details. He was told he would be put in irons (if he resisted, double irons). However, he recognized a friend from his days in the "old navy," Lieutenant Commander W. D. Whiting, the *Ottawa's* commander, who interceded and accepted Glassel's parole. James Sullivan was found the next morning by the *Iron-sides'* crew and was unceremoniously thrown into a dark cell.

Rear Admiral John A. Dahlgren was furious. He warned the prisoners that they would be taken to New York, tried, and probably hung "for using an engine of war not recognized by civilized nations." Privately, the admiral knew this was sheer bravado, and he cautioned Secretary Welles: "It is desirable that this Officer should not be allowed to return here until some time has elapsed, as he could not fail to be of great service to the enemy in future operations of the same kind." [4]

An inventor of great ability himself (the Dahlgren gun was standard Navy equipment) he stated his real feelings in his reports: "Among the many inventions with which I have been familiar, I have seen none which have acted so perfectly at first trial. The secrecy, rapidity of movement, control of direction, and precise explosion indicate, I think, the introduction of the torpedo element as a means of certain warfare. It can be ignored no longer. If 60 pounds of powder, why not 600 pounds?" In his diary he was even more commendatory: "It seems to me that nothing could have been more successful as a first effort. . . ." [5]

The light of day revealed to both sides that the *New Ironsides* was still afloat and apparently undamaged, but this was deceiving. Glassel had hoped to set his torpedo for eight and a half feet; however, a flaw in the spar (an old boiler tube) subtracted a foot from this depth. He and Beauregard always believed that the explosion had been too near the surface to direct its full force on the target. Also, by pure happenstance, the torpedo had exploded

"David" torpedo boat *No. 4* after the capture of Charleston. The torpedo spar can be seen on the bow.

The Confederate submarine *H. L. Hunley,* from a painting by Conrad Wise Chapman.

Rains keg torpedo. This example of the most widely used mine, on land or sea, was found in August, 1863, near Charleston.

Friction torpedo activated by an operator ashore who pulled a lanyard.

Current torpedo designed to float against enemy ships. The propeller, turned by the current, released a spring-driven plunger.

Spar torpedo.

Horological torpedo captured in the St. John's River, February, 1864.

Floating tin torpedo.

Fretwell-Singer torpedo. The plunger was activated when an iron weight was knocked from the top of the air chamber.

Copper swaying torpedo. This device, like the one at left, was anchored with the air chamber up.

Large spar torpedo boat beached at Charleston.

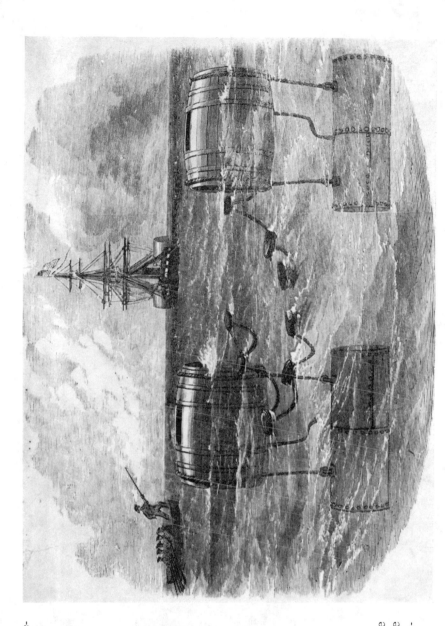

The first torpedoes of the Civil War, used against the U.S.S. *Pawnee* in the Potomac River, July, 1861.

A Confederate torpedo assembly station on the James River, captured in 1864.

The *David* attacks and damages U.S.S. *New Ironsides*, Charleston Harbor.

U.S.S. *Cairo*, sunk in the Yazoo River, December, 1862.

The raising of the pilot house of the U.S.S. *Cairo* from the Yazoo River in 1960.

where a structural bulkhead abutted the hull, and the wall absorbed most of the shock. Still, there were casualties: Ensign Howard died from a single buckshot ("It savors to me of murder," thought Dahlgren.); Seaman William L. Knox had a broken leg; and Thomas Little, master at arms, suffered several contusions.

Before Howard's death, in recognition of his services, the *Ironsides'* commander, Commodore S. C. Rowan (the same man who discovered the first torpedo on the Potomac) recommended that Howard be promoted to acting master. On October 6, Rowan wrote that he had "always behaved well under fire, and was particularly conspicious last night." Dahlgren thought the young man "in every way deserving of the advancement" and passed on the report to Secretary Welles. On the back of the letter is Welles's penciled endorsement that the promotion, "for gallant conduct in face of the enemy," was granted posthumously on October 16.[6]

The big ship was thought to be lightly damaged and a diver reported her hull was hardly dented. (One of these hardy individuals had a harrowing experience during the inspection. While he was crawling under the keel in his rubber suit, the vessel shifted suddenly, pinning one of his arms. Frantically he burrowed through the mud as the immense hull settled over him; he managed to escape just in time.) For a time the *Ironsides* remained at her anchorage, but as coal was removed from the bunkers, signs of damage were found. In late November the ship's carpenter made a thorough investigation on the basis of which Rowan reported that the *Ironsides* was "very seriously injured, and ought to be sent home for repairs as soon as 'tis possible to spare her services here."[7]

She was towed to Port Royal, South Carolina, and then to Philadelphia where she underwent repairs. Her hull and supports had been weakened, a large deck beam had been driven "on end" and a knee was shattered. The damage was kept a profound secret, and the Confederates never knew of it until after the war. The fact that the ship was not seen in action for more than a year gave them some idea of its extent, however.[8]

The audacity and the demonstrated ability of the Confederates at Charleston in attacking and nearly destroying the greatest Union warship had repercussions on both sides. Dahlgren instituted a number of measures in the immediate vicinity to prevent such events' being repeated. Every monitor was anchored at a predetermined position, lookouts were increased, and log booms were placed around each one. Calcium searchlights swept the water, picking out whaleboats paddling between the ships. Aboard these craft were sailors with orders to report any vessel bearing the slightest resemblance to the *David*.

That single attack almost reversed the position of the two navies at Charleston. During daylight the Federals were supreme, guarding the harbor and preventing ships from entering or leaving, but after dark, they gathered for mutual protection, allowing the Confederates the run of the waters in the outer and inner harbors unchallenged.

A Union boat under the charge of Acting Master Mate D. B. Corey was patrolling off Fort Sumter at 9:30 the night of October 8, when he heard the rattle of muskets from the monitor *Catskill*. Hailing the ship, he was told of "some kind of craft going toward the fleet." Corey took a "large boat" without oars under pursuit, but lost the mysterious vessel in the darkness. Two nights later he again reported seeing something which appeared to be about 10 feet long above water, going very fast. It seemed "to extend someways under water by the ripple. . . ." The eerie craft disappeared as before, but a sentry aboard the schooner *Dan Smith* spotted it again. He hailed it, received no reply, and sounded the alarm. The crew grabbed rifles and opened fire, apparently driving the intruder away. It was seen again on October 17, and once more at 3 A.M. on the twentieth when the *Catskill* spotted "a low, dark object" on her starboard bow. She too fired, but claimed no results.

These sightings may well have been the *David*, which was now commanded by Tomb, for the success against the *Ironsides* had elated Beauregard. The torpedo boat's journals have been lost, but not all the Union sightings can be written off as figments of the minds of frightened sailors.[9]

Commanders at other southern ports began to follow Charleston's example and build torpedo boats of their own. Admiral Franklin Buchanan, C.S.N., former commander of the *Virginia*, learned of the attack while he was at Mobile, and began planning torpedo rams himself. In early 1863 a small steamer, fitted with the spar torpedo mechanism, was christened the C.S.S. *Gunnison*.

Her 20-foot spar could be raised and lowered by a windlass, and the torpedo itself was equipped with three fuses. She cruised the inland waters off Mobile Bay, and reports of her reached Admiral Henry K. Thatcher of the Union squadron blockading the port. He warned his ships to "be vigilant" for "torpedoes . . . are entitled to no quarter."

At the same time, and probably on the basis of the same reports, ships off Ship Island below the Mississippi coast were instructed to watch for attacks on the night of November 27, 1863. These warnings were well taken, for Midshipman E. A. Swain, C.S.N., was put in charge of the *Gunnison* on November 9, with orders to "destroy, if possible, the U.S.S. *Colorado* or any other vessel of the blockading squadron. . . ." The *Gunnison* was prepared by her new crew for the attack, but at the last moment her engineers decided it would be the better part of valor to stay within the confines of Mobile Bay, and refused to go.[10]

At Charleston, Captain Lee had returned to his first love, torpedo boats, and was requested to review the subject and give his recommendations. This he did with gusto, working out drawings for a larger, more powerful design. Beauregard agreed with his suggestion to build more boats right away, but the general bemoaned the lack of government cooperation. He offered one alternative, however: Lee could contrive to work with the Southern Torpedo Company on their boats—on his own time.

At the same time, Lee fended off another officer who tried to become involved in the "torpedo business." This man was Captain John Ferguson, who had stated his intention of adapting the Lee spar to carry torpedoes of a different type. Lee felt his own plans had been quite successful and resented Ferguson's intrusions. He did not want the untried torpedo put in ships designed for his own explosive.[11]

While side issues were being resolved, the engineer and his general turned once again to Richmond, this time in hopes of floating a whole fleet of "Davids." Congressman Miles was to make the arguments on the floor of the House, enlisting the aid of that all-powerful body. He was sent a list pointing out the defects of ironclads—including the fact that they were primarily defensive vessels—and contrasting them with the *David* and its offensive characteristics. The enemy cordon could only be broken by attacks using Davids in sufficient numbers. This plan "would raise the Blockade of our Atlantic and Gulf coasts, and enable us to recover the navigation of the Mississippi." The argument bore the signature of General Beauregard and contained one statement which is especially interesting when read from the vantage point of a century later: "Indeed, a few years hence, we will ask ourselves in astonishment, how it was that with such a great discovery, offering such magnificent results, we never applied it to any useful purpose in this contest. . . ." [12]

Instead of bringing the subject before the House, Miles elected to put it before the Secretary of the Navy. Perhaps he felt that in light of the attack on the *New Ironsides*, Mallory would take the Davids more seriously than before. The Secretary was very gracious, freely admitting the deficiences that Beauregard had found in his ironclads, but said that if the Navy had wanted to build ships for offense it would have done so. It was committed to fighting inside the mouths of harbors, not on the open seas. He did not regard the Davids as being unpractical, but straddled the issue by stating that the Navy would welcome a new class of warships once an efficient prototype had been developed. At that time he would "be happy" to help get appropriations for them. The torpedo boats, he added, were not the complete answer, for it was quite easy to surround ships with protective devices that would render them invulnerable to this kind of attack. There was, he admitted, a possibility (as Beauregard had suggested) that larger and swifter torpedo boats could be built overseas if foreign governments would permit it. He was willing to try to have this done, but held out little hope of accomplishing anything.

Miles sent a copy of Mallory's letter to Beauregard, who was somewhat upset with him for taking the matter to Mallory instead of Congress. "I do not suppose," he said, "that Mr. Mallory can possibly admit they [ironclads] are worse than useless, since he is still going on with their construction. But I do believe that Congress ought to interpose its authority in thus allowing Mr. Mallory, or Mr. Anybody Else, to squander our public funds in such a wanton manner . . . I think a committee . . . ought to . . . determine the exact value of the iron-clad gunboats. . . ." [13]

Beauregard then started on a different tack and succeeded in having the Army Engineers contract for a few of the Davids, probably as an extension of their duty of obstructing rivers and harbors. As might be expected, Captain Lee was charged with the program. In deference to the difficulty in obtaining materials and labor, the Engineer Bureau did not specify the number of ships to be built, leaving this to the discretion of the men involved. Lee realized that the engine was the critical element and projected his plans according to the probable supply. W. S. Henery, the machinist whose shop he had employed previously, showed him patterns for some "double engines" of his own design that Lee thought fitted his needs exactly. Lee hoped to make twenty pairs, enough for ten boats, and send them to the various shipyards that would build the hulls. This could be done only if the machine shop dropped all other work—casting anchors for fixed torpedoes, preparing shells for artillery, and building gun carriages. He was authorized to request that Charleston Arsenal release Henery for the new work.

The minute Lee's request reached his desk, Major N. R. Chambliss, commander of the arsenal, saw red. He dispatched a blistering letter to General Beauregard, using such phrases as "interference . . . assumption of authority . . . highly prejudicial to military discipline" to describe Lee's action. He ended with a blunt demand for "orders [to] prevent . . . recurrance." Chambliss was lucky in that Beauregard, instead of reprimanding him, merely refused to accept his endorsement and "suggested" that if he wanted to make his point, he had better withdraw his remarks and couch his wishes in correct military terminology. Undaunted, the major apologized but stood his ground and succeeded in retaining the machine shop for the arsenal's exclusive use.[14]

As the boats were being built, the engineers began to collect their crews and organize the Davids into a single command. They asked Congress to grant "a large percentage" of the value of vessels destroyed or captured to crew members and to permit the shipyards to make direct purchases of such material as was necessary. Lee took full advantage of this latter provision, and even sent cotton to Nassau in exchange for the things he needed.[15]

CHAPTER IX

BENEATH THE SURFACE

On February 1, 1862, the *Scientific American* editorialized: "With all their sources of supply cut off, they [Southerners] are thrown upon their own resources entirely, and, as necessity is the mother of invention, they will develop their latent powers and bring them into use. . . . This is a fact which must be patent to all who think, and nothing is gained by trying to conceal it." Such a realistic appraisal of the influence Southern inventors would have on the course of the war probably came as a shock to many readers, for these men had attained little prominence as yet, and the stories that did filter up North were so fantastic that few took them seriously. The *American* was not fooled. After all, until 1861 there had been little call for inventors in the Southern states with their agricultural economy and rural civilization; northerners had dominated the field from necessity. The North claimed the bulk of the country's machines and factories, and it was there that the people depended upon mechanical things for their livelihood. By

1862, Southerners were put in a somewhat similar position, and the conclusions were obvious.

There was a good bit of inventive activity in the Confederacy, even from the beginning, though most of the effort was concentrated in a few urban centers where machine shops and factories were located. Not surprisingly, especially after the proclamation of blockade, much of the activity was directed toward means of destroying enemy ships. Some inventors chose to dwell on floating or static torpedoes; others thought of methods of placing them against the ships. On this subject too, the *Scientific American* made its prediction: "Cannot a torpedo be invented that will blow to pieces any vessel in the world? is a question which had been asked us more frequently perhaps than any other, and we believe that it will be answered in the affirmative." [1] The *David* offered one solution to the problem of getting the torpedo to its target accurately and without prior detonation; another solution lay beneath the surface of the water—literature was filled with accounts of experimental undersea boats, especially in the United States. David Bushnell nearly sank a British warship in New York Harbor during the Revolution. Robert Fulton's experiments were fairly well known, as were those of Ross Winans in Baltimore just before the Civil War. It was foreseeable that similar ideas would be advanced by persons living in the Confederacy.

Apparently, there were several suggestions along this line. Of the concrete proposals, some were attempts to form combines for the creation of weapons which never materialized. Unquestionably, much of this activity was a direct result of the Confederate government's offer to pay bounties to the owners of any device that succeeded in destroying enemy ships.

On the Union side, several submarines were contemplated. Perhaps the best known was the one built at the Navy Yard in Washington for use against the *Virginia* (*Merrimack*). It was thirty-six feet long and was powered by eight paddles that opened and shut like the leaves of a book. [2] Others that sawed through underwater harbor obstructions and put torpedoes on Confederate ships were also suggested.

The Federals were constantly on the alert for evidences of Confederate submarines, and would take defensive measures at the least suspicion of one. An example was the incident of Ross

Winans' hull—a vessel built by a rather controversial Baltimorian who dabbled with various inventions and had been imprisoned for secessionist sympathies. Before the war, Winans had begun work on a large semisubmerged iron ship which, he believed, offered advantages over those that floated on the surface. She was powered by a novel sideways wheel and was designed for transatlantic service. In 1861, he cut off her ends and put them together, creating a small vessel with all the characteristics of a submarine.

This unfamiliar object threw the Hampton Roads fleet into confusion when she suddenly appeared in its midst on October 19, 1861, under tow of the tug *Ajax*. Far from having hostile intentions, Winans planned to have her taken off Cape Henry at the mouth of Chesapeake Bay, filled with salt water, and transported to Baltimore to determine reaction on her interior surfaces. Flag Officer Louis M. Goldsborough was unaware of this, however, and would not release her on the say-so of her crew, because she "could be so easily converted into an instrument of destruction" by the enemy. Consequently, she remained in the Roads until the Navy Department clarified her intent.[3]

At the same time, word reached Washington of a submarine that the Confederates were building at Richmond, and General McClellan's Secret Service chief, Detective Allan Pinkerton, took prompt action. One of his operatives, Mrs. E. H. Baker who had lived in the Southern capital, was sent there again, ostensibly to visit friends. Arriving at her destination on Sunday, November 24, she looked up a Captain Atwater and managed an invitation to stay in the city as a house guest. In the course of conversation, Mrs. Baker "innocently" asked to see the famous Tredegar Iron Works (where the submarine was being built), and at first Atwater assured her she could do so. On second thought, however, he refused, saying that a secret test was to take place that Monday. Mrs. Baker persisted, and learned that this test was on a weapon "to break up the blockading fleet at the mouth of the James." Finally, she succeeded in persuading the naïve Atwater to let her watch the experiments.

Some ten miles down the river they joined a throng of officers, civilians, and ladies gathered on the shore. A large scow anchored in midstream was the target. About half a mile away floated a strange little craft which began to sink, leaving a large green float on the water's surface. Everyone kept his eyes on the float, which slowly approached the barge, stopped near it, then retreated. (The float, Mrs. Baker learned, was connected to an air hose which allowed the crew to breathe while the submarine was underwater.)

In answer to Mrs. Baker's queries, Atwater is said to have described how two or three men wearing rubber diving suits operated from the vessel after she fastened herself to the ship's bottom by a gigantic suction cup. They screwed the torpedo into the target hull and, when the submarine had moved to a safe distance, pulled a lanyard to the fuse. The captain was interrupted by "a terrific explosion" and the target scow was "lifted bodily out of the water."

This discovery so upset Mrs. Baker that "she could scarcely control her emotions of fear for the safety of the Federal boats in the event of . . . successful operation . . . provided the government was not speedily warned of its existence." She was even more perturbed to learn that what she had seen was merely a working model of a much larger ship (which the obliging Atwater showed her) that was scheduled to be completed in two weeks. The spy wrote down everything she had learned and, a day later, made her way back to Washington via Fredericksburg to warn McClellan and Secretary Welles, who in turn alerted the ships in Hampton Roads.

Some sources say that the submarine attacked the fleet in Hampton Roads on the evening of October 9, 1861, at which time she became entangled in a grapnel and, depending upon which version of the story one wants to accept, either sank or escaped to Norfolk. This writer believes that, rather than the submarine, what the Yankees saw were floating mines set adrift by Robert Minor from the *Patrick Henry*. If Minor's report is valid, the weather was such that no craft so frail as the submarine could have lasted in it for long.[4]

There is a possibility that no such craft ever existed on the James River. Pinkerton's version and a letter in the New York *Herald* on October 12, 1861, are both mistaken as to the date of the supposed attack. In fact, Pinkerton's dates are at variance with those in other sources. On the other hand, a sketch reproduced in

Harper's Weekly for November 2, 1861, and some plans captured from a courier who was trying to cross the Mississippi into Texas in November, 1863, were strikingly similar. Even more disturbing to the Navy was a letter taken with the plans, because it was from a James Jones in Richmond, who had supposedly developed the James River undersea boat. According to the reports of spies, Jones later sent more drawings and built one such ship at Houston, and no less than four at Shreveport.[5]

Submarine construction was by no means limited to the East Coast. In New Orleans, James R. McClintock and Baxter Watson, operators of a steam gauge manufactory, designed a machine to produce Minié balls and, in addition, conceived a warship that would cruise underwater to sink the enemy with torpedoes. They engaged the well-known Leeds Foundry in the "old city" (which, among other things, had cast cannon for the Army) to put their drawings into physical form. The work was done in a cradle at the government shipyard in the fall of 1861. Every day, the builders and some of their backers, a well-to-do sugar broker Horace L. Hunley, H. J. Leovy, John K. Scott, and Robbin R. Barron, would visit the yard to watch their "machine" taking shape. Large sheets of quarter-inch boiler iron were bolted onto an iron frame. The air rang as workmen hammered bolts into countersunk holes to provide as smooth a surface as possible. In a few weeks as the lines of the vessel became clear, some people likened her gracefully swept contour to that of a fish. She measured some twenty feet long, four feet deep, and six feet wide; her sides, keel, and deck were all curved. In place of a superstructure and funnel was a single stubby hatch on the deck, amidships; a ridiculously small propeller capped a shaft at the stern. Projecting like wings from the stem were a pair of iron diving vanes operated from within. There was no pilot house and no engine; two groupings of glass-covered holes arranged in one-foot circles forward served as the former, and cranks operated by two men in the hull turned the screw. Interior fittings were spartan: U-shaped brackets bolted to the floor served as seats for the crank turners, and a magnetic compass mounted on a frame member was the sole instrument; candles provided light when the craft submerged, and a hand-operated pump next to a sea cock emptied the water ballast tank.[6]

Only the builders and owners attended the launching. The vessel's name was an apt one in view of her revolutionary character; she was called the *Pioneer*. As soon as the boat was in the water, her crew prepared her for a short run. Scott took the helm and three crewmen operated the "engine." Later, in one of several practice dives, Scott found that the magnetic compass was practically useless. The needle swung crazily under water, and he had no idea in which direction he was heading. McClintock wrote of the compass: "At times [it] acted so slow, that the Boat would . . . alter her course for one or two minutes, before it would be discovered, thus losing her direct course, and . . . compell the opperator [*sic*] to come to the top of the water, more frequently then [*sic*] he would otherwise."[7] Other compasses were borrowed, but none were completely satisfactory. The *Pioneer*'s skipper finally worked out a method of sighting his bearings before submerging and occasionally rising to porthole depth to check on them visually. (It was not until the development of the gyroscopic compass many years later that this problem was solved.)

The *Pioneer* proved to be tight, and only a few modifications were needed to stop her leaks. It was her offensive armament that needed perfection. She could stay underwater for only a few minutes at a time as there was no reserve air supply. This problem, the inaccurate compass, and the limited power to turn the propeller precluded lengthy underwater cruises. Nevertheless, in March she was given a full test. A barge anchored in Lake Pontchartrain was the target. The *Pioneer*'s weapon was a "magazine of powder" equipped with a sensitive fuse, probably designed and furnished by Captain Beverly Kennon. While the owners watched, she glided slowly toward the barge; Captain Scott took his bearings and submerged. The torpedo struck true and the target vessel was blown "so high that only a few splinters were heard from."[8]

The owners applied for and on March 31, 1862, received from the Confederate government a letter of marque classifying the submarine as a privateer, but the tiny *Pioneer* was never to prove herself in battle or to repay her owners by destroying an enemy ship. Farragut attacked New Orleans before she could be used. As McClintock put it, "the evacuation of New Orleans lost this Boat before our Experiments were completed." But all was not lost, for valuable experience had been gained: "this Boat demonstrated to us the fact that we could Construck [sic] a Boat, that would move at will in any direction desired, and at any distance from the surface." [9]

Loss of the *Pioneer* discouraged but failed to defeat the would-be privateersmen. Determined to realize a profit from their venture, Hunley, McClintock, and Watson fled to Mobile with their plans. There, in the summer of 1862, they contracted with Thomas Parks and Thomas B. Lyons—proprietors of a large foundry engaged in the production of artillery, engines, and other machinery for the government—to build a second sub. The keel was laid on Water Street, and when she was partially finished, she was moved to the foundry for completion.

The new vessel was somewhat larger than her predecessor and was much more sophisticated as to regular equipment. She, too, was built of boiler iron and she similar in shape to the *Pioneer*. Her ends were longer and more streamlined "to make her easy to pass through the water." The single rudder was placed aft, and two sets of vanes at the bow controlled diving.

Manpower had proved to be a poor source of propulsion. The crank turners tired easily in the confinement of the hull since exercise increased their use of the limited oxygen supply. At the machine shop, McClintock and the others sought another source of power. Steam engines were clearly out of the question, for they consumed enormous amounts of fuel, and, more important, they could not be operated underwater. The inventors turned to a new and far more revolutionary development—one that would find wide application in underwater craft of the future—electricity. This science was still in its infancy, but already great strides had been made in putting it to practical use. Perhaps it could be adapted to propel a submarine. The men worked on an electromagnetic engine based on such principles as were known at the time.

"There was much time and money lost," McClintock said of the project, for with the limited resources at their disposal they could not build an engine with sufficient power. Resignedly, the builders fell back on their original crude methods; space was made in the hull for a crew of crankers. Two hatchways fitted with glass coamings for better vision and lighting were added along with a mercury guage to indicate depth. The submarine lived up to her builders' hopes in that she handled easily, but many of the old problems were unsolved. Her speed was limited; inside, the crew was still cramped; and once the ship submerged, the air turned foul in short order. All her surfaces and the men's clothing were constantly wet. A single candle gave a faint flickering light which, when it burned to a pinpoint, told the crew their air was nearly exhausted.

After several trials in the harbor the craft was equipped with a torpedo and towed toward Fort Morgan at the entrance to Mobile Bay to attack Union blockaders. The sea was choppy, and swells broke over her, buffeting her severely. Spray and green water poured into the hatches which were left open for ventilation while she was on the surface. Frantic bailing was hopeless; she settled lower and rolled heavily. Her captain fought with the controls, but the submarine became unmanageable, filled, and sank.[10]

Undeterred, McClintock, Watson, and Hunley commenced work on a new submarine immediately, "taking more pains with the model, and the machinery" and selling shares to several more persons, thus raising a total of fifteen thousand dollars. To save time, they began with a ready-made cylinder that could be modified into a ship—a 25-foot ship's boiler which measured four feet in diameter. It was cut in half lengthwise; each part was strengthened by bars of boiler iron along the interior; and the two halves were then rejoined. After the ends were faired, a tapered bow and stern were bolted to them. The "deck" consisted of a 12-inch strip of iron riveted to the boat, and watertight bulkheads were placed forward and aft. The space between these bulkheads and the ends of the boat served as ballast tanks and was equipped with sea cocks and hand pumps. Under the keel was additional ballast consisting of flat, solid iron castings held in place by T-bolts which could be twisted by special wrenches at the deck if the ballast needed to be dropped.

The diving vanes were connected by a 1¼-inch rod that passed through the forward part of the boat. The single propeller, like the one in the first boat, was connected to a shaft powered by men turning cranks—in this case a total of eight men sitting along the port side. Around the propeller blade was a "fence" that protected it from fouling. Two hatches were cut through the deck about fourteen feet apart. Each was topped with heavy glassed coamings, rubber gaskets, and hinged covers with bolts for tightening. Consideration was given to the air supply, and the designers came up with an iron box on the deck, 1 foot, by 1½ feet, by 4 inches, from which a shaft led to the interior. At either end of the air tube was a 4-foot section of inch-and-a-half piping screwed into elbows. The upper air tube was fitted with a block turned by a key which prevented water from entering it when the craft was under the surface.

There were also several refinements on the interior. All shafts, bolts, and crevices were tightly caulked; wrought iron ladders led to the hatches; and new controls were in evidence. The pilot stood so that he could sight through the coamings; and at his hands were the diving levers, the bow ballast pump control, a mercury gauge, an adjusted and improved compass, and a miniature ship's wheel attached to the rudder. A petty officer occupied the rear hatch and controlled a sea cock and pump for the stern tank.

This much-improved submarine was forty feet long, and had an elliptical cross-section. Much was expected of her.[11] She was not a "pioneer" in the sense of her forerunner, but she was, her builders hoped, a thoroughly practical vessel. Her contours caused the workmen to dub her the "fish boat." [12]

The submarine was floated in the spring of 1863 and was seen cruising around the harbor on test runs. It was soon discovered that she lacked longitudinal stability, but after a few trips the officer at the wheel learned how to compensate for it. Following surface tests, the crew prepared to test her offensive powers.

A copper torpedo containing about ninety pounds of powder was tied to a line from the stern; the hatches were closed, the candles lighted, and the sea cocks opened. Water rushed into the tanks and the little vessel settled until only the hatches were above the surface. The captain and the petty officer closed the cocks and peered through their coamings, ordering the engineers to turn the cranks. As the submarine got under way, the skipper sighted his target, an old barge, and noted the compass bearing. Next, he depressed the vanes and watched the green water rise over the glass. When the depth gauge registered the desired figure, he leveled her off. He saw the target's shadow in the water and carefully passed underneath, then, raising the diving lever with one hand while pumping the ballast with the other, he signaled the petty officer to follow suit. The boat was rudely shaken by a deafening explosion, which rolled her over and made her iron hull ring. Triumphantly, she surfaced. Hatches were opened and the grinning officers waved. The experiment had been a success.[13]

Each time the maneuver was repeated, the crew became more and more proficient as they learned the little boat's peculiarities. But the final dive brought near-disaster. Mobile Bay was chopped up by a strong wind, but higher coamings than on the earlier boats kept the water from slopping inside. However, the wind blew the torpedo along faster than the submarine could pull it. Desperately the pilot fought to prevent its hitting the submarine and, in exasperation, cut the line. This method of delivering the torpedo was clearly unsuitable.

Hunley, who seems to have taken charge of the operation, adapted Captain Lee's spar torpedo, attached a 20-foot spar to the bow, and fixed the explosive to its tip. Instead of passing under the target, the sub would sink only to hatch depth and ram the torpedo into the enemy. When the new method was tried she was buffeted roughly by the explosion, but she survived.

About this same time the group of submariners gave up their hopes of privateering. One contemporary reported that Hunley "admired Genl. Beauregard above all men, and offered him his boat. . . ." But the best reason appears to have been the $100,000 offered by John Frazer & Company of Charleston to anyone who would sink the *New Ironsides* or the *Wabash,* and $50,000 for every monitor sunk. At any rate, the craft was offered to Beauregard, who was quick to accept, and in August, 1863, she was loaded aboard a pair of flatcars and sent to Charleston.[14]

At the beautiful old South Carolina city the submarine became the object of great interest. A request for volunteers to man her brought responses from Lieutenant John Payne, C.S.N., and seven sailors. They supervised the lowering of the boat into the water and received orders to attack the *New Ironsides.* A few practice dives were made to familiarize the crew with the vessel. On the day of the projected attack, they moved her to a wharf at old Fort Johnson on the south side of the bay, and shortly after midnight they cast off and started for the enemy anchorage. The little boat had barely cleared her berth before she began to pitch violently in the wake of a passing steamer. Payne tried to point her into the swells, but the water broke over her and cascaded into the hatches. In a twinkling the submarine settled, her slim buoyancy overcome by the weight of the water inside. Her commander shouted for the crew to abandon her, and dove overboard. Before his shipmates could follow, the submarine sank.[15]

Clearly, navy crews did not handle the sub properly. This, said Commodore Ingraham, was "a New Fangled boat" and required special handling. Personnel with more knowledge of the ship, who fully understood the new elements involved in operating her, were needed. Who was better qualified than her builders? With this thought in mind, Beauregard requested that Mr. Hunley come to Charleston and take charge of the ship. Hunley brought with him Thomas Parks, Lieutenant George E. Dixon of the Twenty-first Alabama Infantry Regiment, W. A. Alexander of the same unit, and several more men who had served the boat earlier. When they reached Charleston, the "fish boat" had been pulled from the harbor and was tied to a wharf. Many conferences were held with those in command of the city, and a new plan of attack was worked out. Dixon was placed in charge of her, and Hunley, Parks, and Alexander were to handle servicing and repairs. The old method of towing the torpedo was revived, but before he would agree to an attack, Hunley demanded that the new crew have plenty of experience.

The submarine became a common sight in the Ashley and Cooper rivers and on the waters of the inner harbor. On one run she passed under the Confederate receiving ship, *Indian Chief,* starting her dive about 250 feet off its port side, and disappearing for more than twenty minutes. Those aboard the ship were becoming quite anxious; then a crewman shouted and pointed past the rail where, some three hundred feet away, the little black hull had broken the surface.

Often Dixon would take the craft down for as long as a half-hour at a time, but he hesitated to try and set a record. On

October 15, 1863, a fine calm day, Dixon was sent out from Charleston on temporary orders, and Hunley ordered the crew to assemble for a practice dive with himself at the controls. As usual, they were to pass under the *Indian Chief*. Hunley had more experience with the submarine and knew her better than anyone else. Some three hundred feet away from the *Chief* he closed the hatches, sighted the ship, marked his compass, and opened the ballast. Evidently, he opened the cock too quickly, for the boat tilted downward sharply and her prow struck bottom, throwing the crew off balance. A subsequent investigation showed that, apparently, the forward ballast tank overflowed and water spilled into the cabin, rising around the men. Hunley grabbed the handle of the pump and worked it feverishly, shouting at Parks to do the same in the rear. The stern began to rise, but regardless of how fast the commander manipulated his pump, the bow remained down—in his panic, Hunley had forgotten to shut off the cock leading to the ballast tank. Desperately, he ordered his crew to release the iron ballast bars under the keel. They grabbed wrenches and fumbled in the water for the bolt heads; before they could turn the keys completely, the rapidly rising water filled the interior of the vessel. As a last effort, Hunley and Parks unscrewed the bolts on the hatches and pushed, but the weight of the water above held the lids down tight. Trapped, the men suffocated.[16]

On board the *Indian Chief* members of the submarine group began consulting their watches. All they had seen were clusters of bubbles that had risen soon after the sub disappeared. Finally, at noon, they accepted the verdict. The "peripatetic coffin" had claimed another crew.

A pall of gloom descended over those who had worked to perfect the submarine. Beauregard canceled the entire project and dispersed the crews. He refused to add more men to the already long casualty list, a list of men killed even before they were within striking distance of the enemy. Usually an enthusiastic supporter of novel schemes, the general washed his hands of the project.

But even this tragedy failed to completely dishearten the remaining submariners. Despite the deaths of their comrades, Dixon and Alexander wanted to raise the vessel and make her ready for use. A diver found her at nine fathoms, at an angle of thirty-five degrees, her bow buried in the mud. The engineers sent a salvage vessel to the spot, and, nine days after she had disappeared, the submarine was placed on a wharf.

With mixed emotions, Alexander, Dixon, Beauregard, and others gathered around the little hull while the hatches were opened. As the iron lids were raised, foul, fetid gas and air escaped. Just below the forward cover stood Hunley's body, its black face bearing an expression of "despair and agony." His right hand was over his head, where it had been pressing against the hatch; in his left was the candle. Parks was found in the after hatch, his hand also pushing the cover. Below, the bodies of the crew were "tightly grappled together." Parks's nearly empty ballast tank had lifted the stern; Hunley's tank was filled, the cock open, the wrench on the bottom of the boat. Bolts holding the iron ballast were partially turned.[17]

Dixon and Alexander began cleaning and repairing the submarine. In the face of Beauregard's repeated refusals to allow them to float her, they employed a different tack and made application through his chief of staff, General Thomas Jordan. They received "many refusals and much dissuasions" before a compromise was reached in which the ship would be fitted with the spar torpedo, used as a David only, and would attack on the surface.

Once again the submariners went aboard the *Indian Chief* in search of a crew. With utter frankness they explained the hazardous nature of the enterprise and the full history of past failures. Probably, they also mentioned Frazer & Company's promise to pay them for any of the enemy's ships they could destroy. More than enough volunteers were "readily procured" wrote Alexander, and a crew was selected from them.

Determined to succeed, Dixon and Alexander embarked on a rigorous course of instruction. They set up America's first submarine school at Mount Pleasant on the north side of the harbor. First, they explained in detail the design and function of the ship; then they undertook cruises. A routine was established: mornings were given over to classroom instruction and exercises; at 1 P.M. the men assembled and marched several miles along the beach to Battery Marshall where the submarine was moored, often coming under the fire of Union artillery across the bay. Aboard the ship, now christened the C.S.S. *H. L. Hunley* in honor of her late commander and builder, a careful check was made of the equipment, after which she was often taken out for short practice cruises, brief dives, and runs along specified courses. At dusk, the officers would take compass bearings on the nearest vessel, "ship up" the torpedo boom and steer for the target ship. On returning, the warhead would be secured, the craft placed under guard, and the crew sent back to its barracks.

November brought choppy seas, but it also brought longer nights. The *Hunley* was capable of about four knots in completely smooth water and light current. Any attack demanded an ebb tide to set out and a flood tide to return; there must be a fair wind and a dark moon. Darkness, especially, was needed, for the new manual of training (contrary to instructions) called for short underwater trips at a depth of six feet, and surfacing every few minutes for air and observation. Darkness would hide the submariners' approach, but there must be enough light for them to distinguish the target.

As winter approached the schedule was modified until the sub averaged four trips a week. The voyages became longer—four, five, six, and seven miles each way. Often when they surfaced near the enemy fleet, the crew heard Federals in picket boats shouting and singing. Occasionally, the *Hunley* was caught up in the tide and it was all that Dixon's men could do to keep from being swept out to sea. Once or twice daybreak found them within cannon range of the Union fleet, but they managed to slip away.

An endurance dive was inevitable. Perhaps if the opportunity to attack had arisen it would not have been, but the constant waiting and the short shallow dives told. One bright, calm, winter day they set out in Back Bay. Every instrument, pump, and fitting was examined, and when all was ready Dixon signaled observers ashore, who marked their watches. He closed the hatches and the craft sank amid a cloud of bubbles. When she gently bumped the bottom, he lighted the candles and trimmed her off. Each man was to turn his crank only a certain number of times to equalize the consumption of air and still allow Dixon to hold the ship steady in the slight tide. The hull sweated as the air became warm and moist; the crew remained motionless and conversed in low tones. The instant any man felt he could endure it no longer, he had only to say "Up!" and the sub would surface. This had the effect of making each man hold out as long as he could, lest he show a sign of weakness. At twenty-five minutes the candle's flame became smaller, then went out. Occasionally, Dixon would ask, "How is it?" Each time the reply "All right" was made by Alexander in the rear hatch. Men gasped and coughed as the air became foul. Visions of the deaths of former crews flashed through their minds.

On shore the watchers studied timepieces and the *Hunley's* last position. One hour passed, an hour and a half, two hours. It was getting dark; no crew had stayed down so long and lived. Regretfully, the observers sent a message to Beauregard: the submarine had claimed another crew.

But beneath the surface all were alive, though gasping. Then, almost in unison, came the cry, "Up!" Dixon and Alexander manipulated the pumps. Slowly, the bow rose—higher and higher it went—but the stern remained on the floor of the bay. Alexander worked his pump frantically, but the angle continued to increase. The men had to clutch the fittings to keep from sliding to the stern and adding to the weight already there. Alexander reached over the top of the ballast tank in desperation and groped for the opening of the sea cock. He was certain that the handle was closed, but perhaps the valve was still open. Nearly breathless, he pulled the cap from the pump and yanked out the valve. In the darkness he felt something wet and slippery clogging it. Seaweed! He cleaned the fitting and rammed it back in place, replaced the pump, and gripped the lever. With agonizing slowness the stern began to rise until the boat leveled and rocked with the waves. The men tore open the hatches and threw back the lids. Years later, Alexander would remember the sensation of that first moment:

"Fresh air! What an experience!"

The crew lay motionless, inhaling the cool, sweet air as it filtered through the craft. Slowly, they became aware of the darkness outside. How long had they been down? The *Hunley* drew up alongside the dock where only a single soldier remained. They tossed him a line, and as realization dawned, he shouted happily. A check was made of the watches, which showed that the

Hunley had been submerged for a total of two hours and thirty-five minutes—a record of which her crew had every right to boast.

The next morning the officers reported to General Beauregard, who congratulated them warmly and expressed his confidence in ship and crew. Gabriel Rains, who was present, expressed disbelief. He was told that they would gladly give him a demonstration—inside the ship. He declined.

Certain that the *Hunley* could be used in attacking the enemy, Beauregard published Special Orders No. 271 on December 14, 1863, after discussing it with Dixon, who, he said,

will take command and direction of the submarine torpedo-boat *H. L. Hunley*, and proceed to-night to the mouth of the harbor, or as far as capacity of the vessel will allow, and will sink and destroy any vessel of the enemy with which he can come in conflict.

All officers of the Confederate Army in this department are commanded, and all naval officers are requested, to give such assistance to Lieutenant Dixon in the discharge of his duties as may be practicable, should he apply therefor.[18]

Accordingly, the submarine did make an attempt, but the enemy's precautions rebuffed it. Every Union ship was surrounded by rafts of heavy logs which were patrolled by picket cutters. The nearest target beyond the harbor was the *Wabash* which was over the bar some twelve miles out. Though the attack was called off, the order stood, to be executed at the first opportunity.

In the meantime the crew continued to train, anxiously awaiting their chance. Among the sources Dixon consulted about conditions were the "old pilots" at Charleston, who knew the vagaries of wind and weather. One day in late January, 1864, he received encouraging information: for the next two weeks the wind was expected to stay in the same quarter, and he could make fairly definite plans. Then, unexpectedly, Alexander was ordered back to Mobile on February 5 to supervise the manufacture of rapid-fire guns at the Parks and Lyons shop.

Dixon had lost a valuable assistant, but rather than delay his plans, he decided to go on without Alexander. At dusk on the peaceful evening of February 17, 1864, with "the bay as smooth as a small pond," the *Hunley* cast off her moorings and sailed into history. Off Beach Inlet, a few miles south of the entrance to Charleston Harbor, the *Housatonic*—a wooden corvette of some 1,800 tons, mounting twenty-three guns—was guarding one of the channels used by blockade runners. The screw sloop *Canandaigua* was hove to out of hailing distance to the southwest, and, within a radius of several miles, their lights pinpointing their positions in the dark water, lay the *Paul Jones* and the *Mary Sanford*. Aboard the *Housatonic* all was quiet. Her officer of the deck, Master J. K. Crosby, was inspecting the securing of quarters. Below decks the rest of the crew was preparing for sleep.

At 8:45 P.M. Crosby noticed a ripple in the water making little eddies and reflecting sparks of light. Peering over the ship's side, he made out what seemed to be a plank, about a hundred yards away, moving toward the ship's starboard quarter. All ships had been alerted to the danger of Confederate torpedo boats, and Crosby shouted a warning. The drummer beat to quarters, and in a moment the *Housatonic* was alive with activity. The anchor chain was slipped, power was applied to the screw, and the ship began moving backwards. A gun crew limbered the after pivot, swinging the muzzle from port, while Captain Charles W. Pickering, his executive officer, and other members of the crew sprayed the as yet unidentified object with rifle-fire.

It was too late. A deafening explosion spewed tons of water into the air just forward of the mizzenmast. The *Housatonic* was lifted from the water and thrown on her side by a gigantic force. Below, seawater and choking fumes filled the interior. Men on deck were flung into bulkheads or over the rails. Guns, shot, timbers, and fittings flew in all directions. As the ship fell back, she settled stern first, heeled violently to port, and sank. Crewmen climbed into the rigging or jumped overboard. Two boats were hastily loaded—so full that water sloshed over the gunwales.

The first that the crew of the *Canandaigua* knew of the disaster was when they were hailed by a small boat. Both loads of survivors soon were taken aboard and the ship went to the assistance of the stricken *Housatonic*. Three rockets were fired and the lights of other vessels moved toward the scene. The *Housatonic* had settled to the bottom by the time help arrived, but the rescue work went on for hours. Five of her crew were never seen again.

In Charleston, Confederate lookouts reported an explosion in the blockading fleet and took note of the unusual activity. The light of day revealed that one ship was missing from her station. Soon afterwards a Union midshipman captured in a picket boat was interrogated, and told what had happened. The reaction in Charleston was one of joy, and on February 24 the Charleston *Daily Courier* reported that the sinking "has raised the hopes of our people, and the most sanguine expectations are now entertained of our being able to raise the siege in a way little dreamed of by the enemy."

But what of the *Hunley?* Only a few people knew she had made the attack. No enemy had reported seeing her or capturing the crew. Were they being held secretly? Had they drowned? A Yankee diver descended to inspect the *Housatonic* but did not mention seeing any other ship on the bottom. The Federals, too, wondered what had happened. They assumed the *Hunley* had escaped and that the Confederates were keeping it hidden. In Mobile, Alexander waited, and when he failed to get any word from Dixon, wired General Tom Jordan repeatedly. He always received the same answer: "No news."

Years later, a diver reportedly found the submarine lying in the sand near her victim. From this discovery Alexander deduced what had happened. He felt that the *Housatonic* had unquestionably backed down upon the *Hunley*. Dixon dove and attacked, but instead of steaming ahead, the warship overrode him and bore him down as she sank.

There is still disagreement as to the number of times the *Hunley* sank and the number of men who drowned in her. Some writers state that she went down once at Mobile and four times at Charleston, taking a total of forty men with her; others have questioned whether another swamping might have taken place under John Payne's command not long after the first disaster at Charleston. Opinion is—and probably always will be—divided. Even the accounts of eyewitnesses do not agree.

After the war, McClintock summed up the saga of the *Hunley* and hinted at the new world of navigation and warfare she had opened:

The Boat and machinery was so very simple, that many persons at first Inspection believed that they could work . . . without Practice

. . . and although I endeavoured to prevent inexperienced persons from going under water . . . I was not always successful. . . . I was . . . willing to . . . show them the difficulties, which in my own mind, was want of speed and Power, I never considered there would be any difficulty in going out and in destroying a vessel, but the Power was not sufficient to bring the Boat back . . . in her destroying of the . . . *Housatonic* . . . they did not work [tow] the torpedo as was contemplated by me.

Since the war, I have thought over the subject considerable, and am satisfied that the Power can easily be obtained . . . to make the submarine Boat the most formidable enemy of Marine warfare ever known. . . .[19]

CHAPTER X

ELECTRIFYING SUCCESS

A lull in stationary torpedo warfare developed during the latter part of 1863. It seems to have been due not to any reluctance on the part of the Confederates but to the Federal ships' reluctance to penetrate farther than the mouths of Southern rivers and harbors. The *John Farron*, a transport, was seriously damaged in the James in late September, but, except for the loss of the *Housatonic*, the Union Navy remained free of torpedo casualties for several months. Besides the nets and booms placed around individual ships and anchorages, there were picket boats cruising the fleets, scooping up and destroying mines in large numbers. Finally, the Yankees began using torpedoes themselves. This brought about a rather interesting series of stalemates, for in the upper James, the Roanoke, and other rivers there appeared two sets of torpedo barriers: one, always the inner, set up by the Confederates to keep the Federals out; the other planted by the Northerners for the opposite reason.

Little had been heard of the electrically controlled torpedoes in the James that Hunter Davidson's Submarine Battery Service had set out in 1862. They were still there, poised and ready for instant action. Their positions were indicated by pairs of sticks set on the river bluffs about ten feet apart and in line with the mines. The operators were to touch off the explosives when enemy vessels came in line with the posts.

In early August, 1863, a Federal expedition probed upriver and passed the first torpedo station. The station's operators were away in Richmond, making more torpedoes with other members of the service. At 4:30 P.M. on August 5, the enemy went back downstream. To the delight of the Confederates, who had returned to their stations, one of the ships passed near the first torpedo. She was the *Commodore Barney*, a former New York ferry, purchased in 1861 and now a 513-ton gunboat which had as a passenger Major General John G. Foster. Excitedly, the torpedo operators set their weapon in operation. Water boiled up under the *Barney*'s bow, followed by a rumbling explosion. The ship's bow lifted and then settled as white foam and chunks of river mud from a geyser nearly fifty feet high crashed onto her deck. "It was terrible," wrote a witness. "The vessel was lifted . . . upward of ten feet out of water, and an immense jet of water was hurled from her bow into the air, falling over and completely deluging her. . . ." [1]

Twenty sailors disappeared, two of whom were drowned, but the *Barney* bobbed up and floated. Excited officers dashed about, bellowing orders, and a damage control party ran forward to check the hull. It was found to be secure, and the ship was in no danger of sinking. Her main damage was a broken steam pipe that disabled the engine. The *Cohasset* took the *Barney* in tow the five miles to Dutch Gap, where her engine was fixed. The next day, the *Commodore Barney* participated in a furious battle with shore batteries and received far more damage (thirty shells struck or glanced off of her) than from the torpedo. She owed her escape not to any action on the part of her crew but to a stroke of fortune. The battery used by the Confederates to ignite their torpedo simply failed to yield sufficient current (so claimed one of the operators after the war), but nevertheless caused an explosion so large that the Federals thought two torpedoes had exploded simultaneously.[2]

There was no further activity until 1864, when a joint Army-Navy expedition commanded by Major General Benjamin F. Butler and Admiral S. P. Lee went into action on May 4. The fleet included the ironclads *Roanoke, Onondaga, Canonicus, Tecumseh, Saugus;* the former C.S.S. *Atlanta;* the gunboats *Dawn, Osceola, Commodore Jones, Stepping Stones;* and several transports. Their mission was to capture Drewry's Bluff, a strongly fortified position above Bermuda Hundred and City Point near the juncture of the James and Appomattox rivers. The increasing successes, the widespread uses to which torpedoes were being put, and the respect with which they were regarded by the Federals were exemplified in Admiral Lee's orders: "All . . . vessels will be fully prepared to drag for torpedoes themselves, and with their boats. . . . at Harrison's Bar, the *General Putnam* and *Stepping Stones* will go ahead and drag . . . carefully . . . taking care to keep 100 or 200 yards apart, so that they do not explode the torpedoes under each other." [3]

None of the ironclads was to move an inch until the water ahead had been cleared, and companies of sailors had explored the enemy-occupied left bank, seeking electric torpedo stations. This slowed the advance literally to a walk, and gave the Confederates all the warning they needed to camouflage their defenses.[4] Under cover of night, the defenders slyly shifted their cumbersome batteries and wires to the swampy right shore and prepared to give the ships a warm reception. At Jones' Neck, where the river narrows and executes a sharp bend, a pair of 2,000-pound tanks of gunpowder had been sunk, and the operators were kept informed of the enemy's progress by telegraphic dispatches via Hunter Davidson's excellent wire network. Copies of the messages were sent to the Navy Department in Richmond where they were anxiously studied by the highest officials, including President Davis.

On May 6, while the ships were coaling off the mouth of Four Mile Creek opposite Jones' Neck, a slave picked up by the shore party reported these electric ambushes. The advance ships, *Commodore Morris* and *Commodore Jones*, were signaled. The *Jones,* in the lead, moved with great caution; her small boats rowed about her, sweeping the channel with their grapnels which actually passed over some of the torpedo wires but did not catch them. At 2 P.M., some five hundred yards from the spot indicated by the contraband (as escaped slaves were then called), she stopped and reversed her paddle wheels to keep from being swept aground. The *Jones* backed directly over a torpedo that had been underwater for twenty-two months.

Ashore, hidden in pits covered with branches and weeds, members of the Submarine Battery Service studied the ship's movements, carefully sighting down the range stakes. Master Peter W. Smith controlled the contacts for the first torpedo which lay in the channel. He lined up the former ferry and joined the contacts.

Suddenly she appeared to be lifted bodily, her wheels revolving in mid-air; persons declared they could see the green . . . of the banks beneath her keel. Then, through her shot to a great height, a fountain of foaming water, followed by a dense column thick with mud. She absolutely crumbled to pieces—dissolved as it were in mid-air, enveloped by the falling spray, mud, water, and smoke. When the turbulence . . . subsided, not a vestige of the high hull remained in sight, except small fragments of her frame which came shooting to the surface.[5]

Forty of the *Jones*'s officers and men disappeared with her. Her engineer, deep in the hull, working the "starting-bar" of the engine, miraculously escaped with a concussion. His only recollection was "a chaotic end to his manipulations," and being picked up by a rescue boat. In Richmond, Lieutenant Hunter Davidson, as a result of the sinking, was promoted to captain by his President "for gallant and meritorious services." [6]

Just before the disaster, a boatload of marines and sailors from the *Mackinaw* had landed on the left bank and were searching a cluster of farm buildings. While dashing back to the cutter after the explosion, they saw a man running along the opposite bank. A volley was fired and he fell, shot through the head. Nearby was a galvanic battery. Two more were found, the third just as its two operators were about to explode a torpedo which was near the wreck of the *Jones*.

Two prisoners captured at the third station were identified as

Peter Smith and Jefferies Johnson, a private. Smith adamantly refused to divulge the locations of other mines, but General Butler placed Lieutenant Commander Homer C. Blake, skipper of the double-ender gunboat *Eutaw,* in charge of Johnson: "If you can use him, do so," he ordered. Blake put Johnson in the bow and told him he would share the fate of the ship.

"He only went about 300 yards," said Blake, "when the man called out: 'Stop, Captain, for God's sake! There's a torpedo just over there!'" He then told all he knew about the mines and enabled the Federals to remove a number of them from the river. One of them held nineteen hundred pounds of powder.[7]

The loss of the *Jones* stopped the fleet's upriver movement and it lay below, at Bermuda Hundred. The Rebels then resorted to other types of mines and began releasing scores of small torpedoes to drift down among the ships each night. Several types were used: tin cylinders with friction primers and floats designed to become entangled with paddle wheels or propellers; copper tanks; and others with slow matches burning toward fifty to one hundred pounds of powder. The Federals scooped up most of them during the daylight, however, and none of the ships were damaged. Also, time was on the side of the Union and the ships struck back at the torpedo stations occasionally.

On July 11, 1864, Cox's Wharf—Davidson's sole remaining station between Chaffin's Bluff and Jones' Neck—was attacked by Company G, Third Pennsylvania Artillery, and a detachment of the Tenth Connecticut Volunteers. Approximately twenty-five men were landed from the *Stepping Stones* at ten that night below Dutch Gap. They marched overland to Cox's Wharf where they destroyed a signal tower, mills, stables, and most of the other outbuildings on the farm, then captured fourteen Rebel soldiers, a horological torpedo, and a galvanic battery.

On May 11 a company of the Eighty-fifth Pennsylvania Infantry had destroyed another station in a similar way. In this case, Confederate infantry was present and put up some resistance before being overwhelmed. Later, while examining the sand at the water's edge, the Yankees stumbled on a "rope." They pulled it and felt a heavy weight, but instead of exploding a torpedo, much to their surprise, the cable broke. The men traced it to a battery hidden in a farmhouse.[8] The expedition was suspended at this point, and the river remained in Confederate hands. The contribution of the electrical torpedoes to the outcome of the engagement did not go unnoticed. "To the use of the submarine batteries placed by you, and others in the James River," President Davis later wrote Davidson, "is most probably due the failure of the Enemy's [sic] fleet to attempt the ascent of the river. A measure so clearly indicated by the movement of their land forces." [9]

During the fall of 1863 there were still a few waterways being actively contested. Among them was the broad, blue St. John's River in northeastern Florida. That torpedoes would appear in this gateway to the interior of the state should not have come as a great surprise to the Federals. They had, in effect, been warned that the Confederates intended to use the devices there, and the first encounter came about as the result of a Union maneuver.

The river received new importance in February, 1864, when General Truman Seymour began an ill-fated campaign to wrestle control of North Florida from the Southerners. The St. John's was vital for transportation and communication, and Union ships patrolled it constantly. The first indication that it would be a contested area came on March 6, 1864, when several warships surprised a party of Confederates trying to launch torpedoes from the banks. The soldiers fled and left two mines of a new type hidden in the bushes near the Sisters Creek. Later, two more were discovered, but how many had been floated was still unknown.

Like the horological torpedoes the *Essex* had recovered on the Mississippi in February, these mines had a clockwork mechanism that rendered the explosives harmless until a spring released a hammer at a predetermined time. They were credited to the ingenuity of Thomas L. Buckman, the C.S. Army ordnance officer for Florida and former superintendent of the Florida, Atlantic and Gulf Central Railroad. Filled with seventy-five pounds of powder each, they had been set to explode an hour and forty minutes after they were found. [10] The Yankees believed that the torpedoes were used in hopes of opening the blockade of the St. John's so that the Confederate steamer *St. Mary* could make a run for the sea. When this plan failed, the Rebels resorted to defensive mines.

A Federal garrison, established at the town of Palatka a good distance upstream, had increased river traffic to keep the soldiers supplied and reinforced. When this lifeline became the target of Rebel efforts, Beauregard sent Captain E. Pliny Bryan from Charleston to take charge of the torpedo defenses in Florida. His first requisition—for twelve forty-pound keg torpedoes, eight hundred pounds of powder, three hundred Rains primers, and anchors—was filled by Captain M. Martin Gray at Charleston on March 12, 1864. Bryan planted a number of the devices near Mandarin Point, a few miles from Jacksonville on the east shore, during the night of March 30. At 4 A.M. on April 1, three Union transports, the *Maple Leaf,* the *General Hunter,* and the *Harriett A. Weed,* were sighted steaming down the river. They had finished debarking troops at Palatka and were going back to Jacksonville for more. Without warning, the *Maple Leaf,* a 508-ton steamer, was jolted by an explosion and quickly settled to the bottom at Beauclerc's Bluff, twelve miles from Jacksonville. She took four men and the baggage of three regiments (the 112th and 169th New York, and the 13th Indiana) with her. Only the twin stacks and "uppers" remained visible. Bryan sent word of the sinking to Major General Patton Anderson, commander of the District of Florida, but could not board the wreck because of foul weather. Later that day a gunboat nosed up to it and made an inspection, departing the following evening.

Meanwhile, a company of the First Georgia Regular Infantry and a section of the Florida Light Artillery went down to destroy what remained of the transport. They reached the riverbank just before daylight on April 2 and unlimbered their brass 12-pounder howitzer, a pair of 6-pounder smoothbore cannon, and an iron 3-inch rifle to command the wreck. After firing a few rounds to give the enemy "the idea that a battery is located there," Bryan and two soldiers boarded the *Maple Leaf.* Water was two feet deep on the upper deck, and in the cabins were "a few mattresses, sofas, washbowls, and other unimportant articles." The baggage, which would have been quite valuable to the Confederates, was unobtainable as it was stored in the hold. Using the mattresses for combustibles, Bryan burned what was left of the wreck.[11]

At 9 A.M. on April 16, the *Cosmopolitan* and the *General Hunter,* returning from Picolata with quartermaster stores and convoyed by the screw gunboat *Norwich,* cautiously approached the wreck of the *Maple Leaf.* Since her destruction, all shipping had moved with care in single file. The *Norwich* led, followed by the *Cosmopolitan* and the *Hunter*. The last ship was having difficulty in staying in the track, for a fresh wind kept pushing her off course. She veered near the wreckage and was rocked by the explosion of a torpedo which "blew the forward part of the hull to fragments." She sank in three minutes. One man, the quartermaster, was drowned at the wheel and the steward suffered a broken leg, but the rest of the crew escaped.[12] Intensive sweeps of this part of the river were instituted by the Federals to rid it of the new menace, but any torpedoes they found were speedily replaced by Bryan with a new supply sent from Charleston.

The next ship to feel their effect was the 290-ton transport *Harriett A. Weed* which, on May 9, struck a torpedo near the mouth of Cedar Creek close to Jacksonville, killing five men. The *Weed* had two guns and was towing a schooner when she was "blown into fragments." Aboard were thirteen officers and twenty enlisted men of the Third U.S. Colored Regiment, all of whom were "more or less injured." One of the officers was thrown twenty feet in the air. When he was pulled from the water, the *Weed's* skipper, suffering from shock, was interviewed by Brigadier General George Henry Gordon, U.S.A. His features covered with coal dust, the unfortunate man wrung his hands and moaned: "Who will come next? How are we to navigate these waters?" Gordon cynically reported that they had no choice but "to take our chances, with the pleasant feeling that at any moment we might find ourselves blown high in the air. . . ."[13]

A pro-Union newspaper editor viewed the situation caustically. When news of the *Hunter's* loss arrived, he commented acidly: "If the United States desire to pay a round sum for pasteboard boats, and to have them used up at an early day, let them duplicate the *General Hunter*."[14]

The Federals were fortunate not to lose more ships in the St. John's, for these sinkings encouraged the Confederates to place large numbers of torpedoes in the river. On May 11, 1864, Charles D. Boutelle of the United States Coast Survey was cruising down the river on the U.S.S. *Vixen*, when, near the *Weed* wreck, he saw a series of "little rips" on the water's surface, indicating objects not far underwater. Robert Platt, a comrade, suggested that possibly they marked the location of torpedoes. Two cutters, sent to investigate, hooked a keg torpedo within an hour. They fired rifles at it, hoping to wet the powder, but failed. Platt tied a knife to a long pole, cut the moorings, and towed the cask to the *Vixen*. There they tried to get it on board, but were hindered by a choppy sea that tossed the cutter and the ship about, increasing the possibility of accidentally crushing one of the torpedo's primers. Platt then towed the mine ashore and bored holes in it with an auger, taking care to pour water over the auger during the operation. The keg was taken on board the *Vixen* to be sent to Washington, and the locations of at least five more mines were pointed out to Commander George Balch of the *Pawnee*.

In early June the Rebels struck again, and, once more, careful attention prevented further losses. Rumors that "loads of torpedoes" had been set in the river off Doctor's Lake opposite Beauclerc's Bluff reached the Federals. They also learned that homes on the riverbank were being used to spy out the locations of both gun and picket boats. An expedition of some two hundred soldiers was sent to investigate, "scour" the land, and remove all loyal inhabitants to the east side of the river. At the same time, boats of all sizes dragged the water. At least seven torpedoes were found, and the Federals threatened to "deal summarily on any caught putting down torpedoes in the river." [15]

CHAPTER XI

PERILOUS WATERS

At Charleston the year 1864 commenced with a Rebel plan to drive a Federal ship onto the torpedoes. Under the direction of Gabriel Rains, torpedoes were placed in the Stono near Legareville on January 10, and at the same time, a "sham battery" was built at Grimball, across the river, to demonstrate against the enemy and to "induce the Pawnee or some other boat to pass through the opening in the piling . . . so she can shell the . . . battery." It was hoped that in so doing she would come into contact with the mines.[1]

Things began to go wrong from the beginning. Rains was sent a unit from Company A, Twenty-first South Carolina Infantry under Engineer Lieutenant J. T. E. Andrews, who had originally thought up the plan. They returned not long after they had set out. "I deemed it impracticable to operate, and returned to 'Battery Pringle' with the determination of prosecuting the work the next night," reported Andrews. "Upon making my intention known to the men, I learned that a majority of them would refuse to accompany me, intimating at the same time that it was my intention to surrender them to the Enemy."

Rains recalled the detail and took the matter up with Captain Reed, the commander of Company A. Reed then made it his "particular request" that he be allowed to furnish another detail, "for whose fidelity and alacrity . . . he undertakes to vouch." Captain Reed's confidence was misplaced, for just before the ship could be lured into the trap, a "traitorous scoundrel" slipped from Battery Pringle and tipped off the Federals.[2]

The deserter merely confirmed what those aboard the *Pawnee* already suspected, that the battery was part of a trap. They had noticed the construction with interest, and the ship actually fell in with the plan when it crossed a gap in the piling over the torpedoes to investigate. However, the sailors became suspicious for two reasons: Confederate soldiers building the emplacements exhibited a strange indifference to the *Pawnee*'s presence and, though she was within range, no shots were fired at her. On January 11 she slipped back to safety, without exploding the torpedoes. Perhaps the tide was too high.

The deserter explained to the Yankees that the ship had actually passed over the torpedoes: "We thought . . . [an explosion] would be the case as you came up and passed through. We were quite sure of it when you returned." The Federals then returned and fished up several mines in plain sight of the Rebels. One of the kegs was sent to the United States Military Academy by Commander George B. Balch of the *Pawnee* and Brigadier General George H. Gordon, a passenger, so that it might "remain a lasting testimonial of the contribution of the traitor Peter G. T. Beauregard to the country for his education. . . ."[3]

On February 4 a band of eight deserters told the Yankees that two large electric torpedoes of fifteen hundred and two thousand pounds had been put on the bottom of the main channel off Fort Sumter, and that smaller ones were anchored two and three fathoms beneath the surface. But these machines represented only a fraction of the many torpedoes that were being sent to Charleston.

A few days later Beauregard instructed Rains to put floating torpedoes in Charleston's main channel near Morris Island and in the Ashepoo, the Combahee, and the Stono rivers, but Rains was detailed to Mobile before he could do all this. Placed in charge of the Charleston "Sub-Marine Corps" were Major W. H. Nichols and Captain M. Martin Gray. Details of the ensuing activities have fortunately been preserved in a daily journal kept by Gray.[4]

On February 20 the captain reported that all orders for the Stono had been filled and that the torpedoes for the Combahee and Ashepoo would soon be ready. Three days later he sent thirty small and a dozen large brass submarine fuses, plus three hundred sensitive primers, by train to Captain William A. James at Wilmington, North Carolina. Seven men were dispatched some fifty miles south to Pocotaglio on March 1 with four more torpedoes to be put in the Combahee and, at the same time, several boxes of parts were sent up to the James River defenses.

The mine operators were to report to their headquarters at 8 A.M. and 3 P.M. each day, and the boatmen and others at 7:30 A.M. and 1:30 P.M. to allow them to prepare for whatever missions had been planned. The foreman could release no more than two men at a time on leave, and there was a general inspection every Saturday morning. Two steamers, the *Molly Myers* and the *Mary Gray,* were assigned to the command to tow the boats as close as possible to the area which was to be mined.[5]

The laying of torpedoes had become a routine operation. There were a total of forty-three men in the corps in January, 1864. Twenty-two were members of the armed services; the rest were civilians, though there was apparently no distinction in their duties according to rank. There were an enlisted clerk and seven mine operators in Charleston (two of whom were civilians), and one of each at Fort Sumter, Savannah, and with the Army of Tennessee. In addition the corps included a rigger, a pilot and coxswain, two carpenters, and twenty-two boatmen—military and nonmilitary intermixed. Supervising their activities were half a dozen officers, ranging in grade from second lieutenant to captain, all of whom were engineers. In charge over all was General Rains. These people were distinct and separate from the crews and handlers of the Davids.

At the corps's disposal were a manufactory, several rowboats, and various military units that were assigned from time to time to protect them during an operation. The boat crews had been specially trained for the work and wore dark clothing and caps when on duty. Their oars were muffled with gunnysacks, and they knew how to cut the blade in the water without splashing. There was, of course, no smoking, but cheeks bulging with tobacco quids were common. Various special equipment was carried in the boats. The crews wore side arms (revolvers and cutlasses) and often carried sawed-off shotguns. There was a plentiful supply of rope, grappling hooks, and several covered tin cans of fresh water in each boat. This last was a thought of Captain Gray's: "The cans being larger than ordinary canteens are especially adaptable for use of men employed in setting torpedoes, being out for a length of time without water." Before the laying of mines began, the route and target were closely studied. From vantage points on the shore the officers would examine the approaches through spy-glasses, look at the charts, figure moon and tide schedules, and have the barometer checked for weather conditions.[6]

After his orders were filled, Gray found time to examine the Stono obstructions (which he asserted to be "in good condition") and to hire nine new men. On March 19 Gray reported that he was working on more electric torpedoes for Charleston Harbor and asked for the loan of a David a week later so that he could plant smaller ones. Captain E. Pliny Bryan, on the St. John's River in Florida, was sent another dozen April 7, and sixteen more several days later. To General Dabney H. Maury at Mobile, Gray forwarded two miles of electric submarine cable which had been obtained through the blockade and, a week later, sent out several officers to round up barrels "suitable for torpedoes" from nearby cities. The fragment of Gray's journal ends with the note that J. S. McDaniel and Captain George H. Proung were at work plotting the positions of the blockading ships, hoping to "destroy them with torpedoes."

The torpedoes in the Stono kept washing away and required periodic replacement. Instead of operating on his own initiative, Gray waited for orders before renewing the defenses. For instance, on April 16, Brigadier General William B. Taliaferro had to ask Beauregard to put out more torpedoes. Confederates at James Island south of Charleston Harbor were expecting an enemy movement in mid-May, having been forewarned by increased Union activity, picket clashes, and shelling from U.S. warships. At 8 P.M. on May 23, Taliaferro asked that more explosives be floated in the Stono, since boats had been seen depositing Federal troops on James Island. As soon as his telegram was received by Beauregard, the Creole sent a party with torpedoes to the threatened area, and, by dawn, four had been anchored in the Stono. Others were being sent from the factory. These precautions did not prevent the enemy from making good his lodgment on James Island, however.[7]

On the other side, Admiral Dahlgren respected both the persons engaged in torpedo warfare and the weapons themselves quite highly, and in May he began using torpedoes. His first idea was to employ sail-powered torpedoes, which helped to clear obstructions, but Brigadier General Alexander Schimmelfennig did him one better and tried a device run by clockwork. It was a failure. Another, developed by a Frenchman, Monsieur Maillefret, did not "work well." On July 3, Dahlgren tried a mine of his own. He told of it in his diary: "The officer rather carelessly placed it only half a length from my steamer. At the word, away it went. The shock was tremendous, breaking all our loose glass. The diameter of the column of water must have been 100 feet, but not very high."

He then set about devising a "mine raft." These large rafts were loaded with several hundred pounds of powder and were to be exploded against Fort Sumter "until the wall is shaken down and the surrounding obstacles are entirely blown away." A test with a hundred pounds of powder went well, throwing up a satisfyingly impressive tower of water—so much water that Dahlgren was drenched while standing on his barge. Twice he tried to float the rafts against the fort, and both times they blew up harmlessly. Consequently, he was forced to content himself with defending his ships against Rebel torpedoes rather than using similar weapons offensively.[8]

Meanwhile the Confederates continued to build torpedo boats. In late January, 1864, the Southern Torpedo Company launched two in Charleston, another pair was being built in Wilmington, and the chief engineer of the Department of the Gulf at Mobile laid down several in his yards. During the spring the War Department ordered Captain Francis D. Lee to build a pair in Charleston, one was started at Selma, Alabama, and still another was begun in Savannah.

Satisfied that a future fleet was in the making, the torpedo boatmen at Charleston decided to make use of those vessels already in the water, reasoning that the earlier they were placed in action, the more ships the government would eventually authorize. Unfortunately, as has happened so often in warfare, the new weapons appeared too soon and in too few numbers, doing little more than alert the enemy, enabling him to adopt defenses.

Though he had only three boats at his disposal, Beauregard was anxious to send them out against the Federals. He had begun using the *David* to lay floating torpedoes as early as January, but this use was in a defensive capacity. A number of choices presented themselves. The *Pawnee* and the *Marblehead* lay in the mouth of the Stono, the *Memphis* was in the North Edisto, and Federal barges shuttled troops and supplies across Schooner's Creek.[9] Chief Engineer Tomb, who was now attached to the blockade runner *Juno* in the harbor, was put in command of the *David* for the attack. He chose the *Memphis* for his target, and set out on the evening of March 4, with a crew of three (two pilots and a fireman), while a battery of artillery moved overland to aid him with a covering fire. The *David* had been fitted with a new spar attachment which allowed the weapon to be lowered to any depth, and had also received an extra layer of iron sheathing and a "cutwater" shield over the bow. These improvements were designed to avert the mishaps of the *Ironsides* attack: insufficient torpedo depth and partial flooding of the torpedo boat.

Stealthily the tiny boat glided toward the *Memphis* until the warship's lights were clearly visible. Just as he was ready, Tomb found that the air pumps controlling the admission of water in the *David*'s ballast tanks were defective. He canceled the project and returned to Church Flat for repairs. The following night, at nearly the same place, the pumps failed him again. Repairs were completed during the day while the little boat lay concealed in the marsh, and she moved toward open water again on the night of March 6. Some fifty yards from the target's port quarter, the *David* was discovered and hailed by the *Memphis*. Tomb refused to reply as beat to quarters rolled across the water. A hot small-arms fire glanced from the shield while he made the run, but the torpedo struck squarely home eight feet under the surface. The *David* stopped for a moment and glided off as those on deck shot at "what looked like a ship's boat . . . bottom up."

The torpedo failed to explode.

The audacious Tomb unhesitatingly spun the wheel to pass beneath the stern of the now-moving warship. Suddenly the *David* shuddered and fell back from a strong blow accompanied by the screeching and tearing of metal. Looking up, Tomb discovered that the boat's stack had struck the *Memphis'* fantail and had been ripped from its sockets. Part of the stack clung to the Federal. Rounding the stern, Tomb smashed the torpedo into the large ship once more. Again there was no explosion. Ignoring a hail of bullets, Engineer Tomb made for the safety of the marsh in disgust. At Church Flat in the cold light of dawn he examined the torpedo and found that the first blow had smashed a fuse but the second had only dented one. On learning of the attack Captain Lee said heatedly, "I distinctly told him, that the torpedo . . . couldn't be relied on, it having been exposed for the last six months to every vicissitude of weather and climate. . . ." [10]

The building program progressed steadily, giving the Confederates cause for satisfaction, while Union authorities grew increasingly alarmed as intelligence reports of the new boats filtered in to them. One of the new vessels sank at Chisolm's Wharf in Charleston soon after its launching but was speedily repaired; the two at Wilmington were christened C.S.S. *Equator* and *Adkin*. Another there (said to be 150 feet long) was under way, while at Richmond one was built on the *David's* lines which became known as the *Squib* or *Infantia*. Nervous Yankees reported sightings in nearly every river along the coast. All the while, the Confederates sought ideas to best employ those they had—including such plans as moving the craft by railroad to various rivers in Virginia and Florida. [11]

Though Beauregard had only four of the thirty boats with which his enemy credited him, he was busy planning an attack using all four simultaneously. This foray began with a report from a reconnaissance party which visited the mouth of the Ashepoo River on March 14. There it spotted two targets, the *Wabash* and a pilot boat, which had been anchored in the same spots for months. Three torpedo boats (two army boats plus the Navy's

David) ran out of Charleston through various streams to the river on April 6. The military vessels were commanded by Captains Augustus Duqucron and E. R. Mackey, the *David* by Tomb. At Royall's House on April 8 both army ships gave out and had to turn back. Alone, the *David* pressed the attack, and at 9:45 that night she was seen by *Wabash's* lookout. She circled her target like an animal of prey as shells bounced from the water. The plucky Tomb finally decided the sea was running too heavily and withdrew.[12]

Union fears of attack in other waters were well founded, for torpedo-boat warfare was beginning to spread like ripples from a stone cast into a pond. It had begun in Charleston, radiated along various Southern rivers, and was about to explode in that greatest (and presumably safest) Union anchorage in Southern waters, the magnificent harbor at Hampton Roads, Virginia.

The Federals had expected such an attempt and had been making preparations for it. During February and March, the Hampton Roads squadron had stationed its lighter ships "to make it highly probable that any attempt on the part of the enemy to approach . . . with torpedoes will be promptly discovered and checked." [13]

John M. Batten, a medical officer fresh from civilian life, bedded down uncertainly in a cabin aboard the mighty U.S.S. *Minnesota* (flagship of Admiral S. P. Lee's North Atlantic Blockading Squadron) for his first night's sleep on the water. About one o'clock on April 9, 1864, he was rudely awakened by a loud noise. "I could not for the life of me tell from where it came or whither it had gone . . . it made the vessel tremble." Alarmed, he threw on his clothes and dashed up on deck. There Batten found the admiral and a cluster of officers and men. They knew the source of the noise: a large torpedo had exploded under the proud ship while she was anchored in the very middle of the squadron.[14]

The attacker was Hunter Davidson. In the laboratory near Richmond, the Submarine Battery Service, learning of the torpedo rams being designed by Captain Lee, had begun its own experiments with spar torpedoes, concentrating at first on warheads exploded by electricity. This plan was abandoned in favor of percussion torpedoes designed in Russia, after the difficulties of stowing the cumbersome batteries inside the hulls of small steamers and of closing the circuit the instant the torpedo came into contact with enemy ships proved greater than any possible benefits. The Russian-type torpedoes exploded on contact and were quite similar to the Lee torpedo.

In the first experiment a fuse was screwed into an empty case which was then put on a spar attached to the *Squib*. She rammed an old wharf at Rocketts but succeeded only in denting the torpedo and shaking up Davidson and his crew. Twenty-five pounds of powder were poured into the case and Davidson tried again, lowering the torpedo two feet under the water's surface. The resulting explosion nearly swamped the *Squib* as water from a sizable geyser cascaded over her; when it subsided, the men saw the wharf was a shambles.

Davidson was ready to attack the enemy in their anchorage at Hampton Roads and lacked only smokeless and spark-free anthracite coal for his boat. There was none in Richmond, but an alternative was suggested: some prewar river steamers that had coaled on the waterfront had used anthracite; surely some had spilled overboard and now lay on the bottom. Indeed it had, and though the project was costly, enough coal was fished from the river bottom to fill the bunkers.[15]

The *Squib* was somewhat different in design from the *David*. Essentially she was a forty-odd-foot steam launch, covered with iron, with a steersman cockpit at the stern and a torpedo windlass on the bow. Commanded by Davidson with a crew of six and armed with a 53-pound spar torpedo, the *Squib* was towed from the upper reaches of the James. Traveling by night, hiding in creeks by day, she traversed the hundred miles to a place near the Federal anchorage without incident. On the night of April 9, she slipped by Newport News and passed through nearly the whole fleet. She had, on occasion, been used as a flag-of-truce boat, and though hailed several times, she did not create an alarm. The position of the *Minnesota* was known, and about 2 A.M. Davidson began his run. The flagship's consort—an armed tug called the *Poppy*, which had the job of circling the larger ship—was motionless when her deck guard saw the intruder. He shouted the challenge several times, only after he had been alerted by the flagship's officer of the deck, Ensign James Bartwistle. Davidson impudently replied that this ship was the U.S. ironclad *Roanoke* and bore in. Bartwistle yelled himself, ordering the stranger to stay clear or be fired on. "Aye, Aye," was the reply, but the *Squib* did not change her course. The *Poppy's* guard began peppering the Confederate with rifle fire. Bartwistle roused his captain and ran to a nearby cannon, telling the *Poppy* to keep the stranger away. But before he could fire his gun, the torpedo exploded "just abaft the port main channel," severely shaking the *Minnesota*. Her crew, led by Captain J. H. Upshur (who had been a classmate of Hunter Davidson at Annapolis), poured out on deck; the drum rattled, commands sounded, and guns unlimbered and fired at the fast-disappearing Confederate vessel.

On board the *Squib*, Davidson's crew were wrestling with the engine which had chosen this moment to "catch on center." Luckily it was fixed by the engineer (who was promoted two grades for his efficiency), and the boat made off as bullets rang against her armor. Hardly a square foot was untouched; one shell splashed under her keel and lifted the stern out of the water. Minié balls zipped through Davidson's hat and clothing, but no one was injured. To throw the pursuers off his track, he started as if to go to Norfolk, reversed his course, and escaped up the James, passing the mouth of the Nansemond River.

On the eleventh, Davidson, who was at Turkey Island up the James, sent a jubilant wire to Secretary Mallory: "Passed through the Federal fleet off Newport News and exploded 53 pounds of powder against the side of the flagship *Minnesota*. . . . She has not sunk, and I have no means of telling the injury. . . . My boat and party escaped without loss under the fire of her heavy guns and musketry. . . ." When he arrived in Richmond he was presented to President Davis. He was shocked when the Executive (who was suffering from one of his recurrent headaches) showed absolutely no enthusiasm and queried grumpily, "Why didn't he blow her up?" [16] Davis soon changed his mind, however, and joined Mallory in proposing that Davidson and his engineer be promoted "for gallant and meritorious conduct." Davidson was made commander when Congress approved the measure in June, 1864.

The *Minnesota* was hurt, not badly, but sufficiently. Beams were shattered, bulkheads sprung, broken hull plates blown inward, and several cannon disabled; but the fact that she still floated did not lessen the shock with which the North received news of the attack. "A little more practice will make them perfect," commented the *Scientific American.*[17] The *Poppy* had been unable to move because of low steam, and her commander was removed and reduced in grade for his laxity.

Rear Admiral S. P. Lee, who happened to be aboard the *Minnesota* during the incident, took a dim view of the attack. He sent word to Davidson that if the torpedo boat was ever used as a flag-of-truce boat again, she would be fired upon, for he did not consider such craft as "engaged in civilized or legitimate warfare," whatever their mission. This did not seem to upset the new commander. It "glanced from my armor as many a worse shot did from my *own side*," he said, "for I felt that as he was the . . . sufferer. . . . he saw the matter from but one point of view, but that time would set it even as I replied in substance . . . '*respice finem!*' "[18]

A search was organized the next morning by the Federals, who hoped to be able to catch the torpedo boat while she was hidden in one of the many creeks along the James. A report that she had run into Pagan Creek, leading to Smithfield from the river's south bank, was checked on April 14 by an expedition led by 35-year-old Master Charles B. Wilder, executive officer of the *Minnesota*. He was in a small boat which had been fitted with a howitzer in case the *Squib* put up resistance when she was found. All was quiet as they ascended the creek with Wilder standing in the bow, closely examining the banks. Then the stillness was broken by the report of several rifles firing from thickets, and Minié balls plunked the water around the boat. Wilder directed the howitzer to the vicinity, fired two rounds, then commenced reloading. Just as he was pulling the lanyard the third time, he dropped to the bottom of the boat, dead from a wound in the left temple. A crewman, H. H. Miller, was severely wounded as the men rowed frantically out of range.[19]

The attack was but a partial success and resulted in an increase in Federal defensive measures. It prompted the South to lay down even more torpedo boats, for the affair had given the Confederate Navy cause for greater hope. Soon afterwards, four of the boats were begun at Richmond.

At Charleston, three of the Army's David torpedo boats were

given orders to move into tributary streams "to attack the enemy's fleet at any place that Captain . . . Stoney may direct." They hovered for a while in the Stono "to operate . . . upon the first favorable occasion" and made repeated appearances throughout the summer. Union shipping was constantly on the alert with instructions to "scour" the water around them at night with grapeshot in case a torpedo boat was near. These precautions proved effective, for there was no foray during this period even though the attacking force had swollen to twice its original size.[20]

In the James, the *Squib* was still a danger and a special Torpedo and Picket Division was organized to search for her. The ships and boats of this command were instructed to "run down the torpedo craft," but none ever had the opportunity. They carefully interviewed refugees and contrabands who entered the Union lines. One reported that "they [the Confederates] say you haven't the sense to make a good torpedo; they reckon on them more than all else besides." On June 1 Admiral Lee telegraphed Gideon Welles that he believed "the enemy meditate an immediate attack upon this fleet with fire rafts, torpedo vessels, gunboats, and ironclads, all of which carry torpedoes. . . . I have not here, and am unable to fix torpedoes which are at all reliable. . . ." This latter remark characterized most Federal attempts at producing effective torpedoes and mines.[21]

CHAPTER XII

THE SECRET SERVICE CORPS

August 9, 1864, dawned hot and sultry over the flats along the broad, muddy James River. Union gunboats, transports, barges, and iron monitors—the latter painted a dazzling white to repel the sun—churned the waters to and from the great supply base General Ulysses S. Grant had established at City Point, Virginia, where the Appomattox emptied into the James, a bare dozen miles northeast of the siege lines around Petersburg. Soldiers, sailors, contrabands, and civilians toiled at the wharves and warehouses that crowded the riverbank. Storeyards were filled with cannon, shells, wagons, provisions, saddles, bales of uniforms, and other paraphernalia of war waiting to be loaded aboard trains and delivered to the armies. Upriver, at Dutch Gap, a line of warships protected the vital base from great, squat Confederate ironclads which occasionally moved as if to attack.

Captain Morris Schaff, U.S.A., who was responsible for the operation of the depot, was preparing to visit the main wharf. He paused at the door of his office atop the bluff to watch the bustle below and saw an old friend, Captain A. W. Evans, coming up the path to join him. During the good-natured banter which followed, Schaff offered his guest a drink but, on inspection, found his supply depleted. "Come over to Grant's headquarters," he invited, "I know where I can get some."

In the tents they found a dozen officers from the staffs of Grant and General George G. Meade gathered around a pail which was two-thirds full of claret punch. The newcomers were quickly supplied with tin mess cups and invited to join the conviviality. Commander John M. B. Clitz, U.S.N., whose ship, the *Osceola*, was anchored in midstream, challenged Schaff and others to a round of "seven up," a favorite card game. They played on a camp bed. Schaff jubilantly captured a pair of tens with a queen, and Clitz, whose turn was next, complacently sat with a full hand, ready to spring it on his opponents.[1]

In the commanding general's log office the nervous provost marshal of the Army of the Potomac, George Henry Sharpe, was telling Grant of a report that Rebel spies were in the area and of the plans he had made to round them up. City Point was especially vulnerable, and strenuous measures were being taken to catch torpedoes that the Confederates had been floating down the river. Only one of these instruments need slip through and bounce against the piling of a wharf or hit an ammunition barge, and the results would be disastrous.[2]

In a dirty, ragged Sibley tent perched precariously on the bluffs near the Negro laborers' quarters, a dusky barber was plying his trade while a line of customers waited. Off to one side, A. M. Baxter of Cold Spring, New York, was busily dispensing lemonade and other soft beverages to a thirsty crowd, happy with the warm weather and his volume of business.[3] Down at the wharf a Confederate soldier—just released from a Yankee prison after seven months—was eagerly pacing, waiting for a vessel to cast off. It was to take him upriver where he would be exchanged for a Northerner who had been captured by the Confederates. A steam whistle sent most of the workers to lunch, leaving only a skeleton crew at work; in an instant this orderly scene was reduced to utter chaos.

First came a muffled explosion, then a cloud of black smoke, and almost immediately the entire depot seemed to erupt in one of the greatest explosions of the Civil War. A huge cone rose from the earth, spreading into a mushroom cloud that was seen for miles. The noise, too, was heard long distances, even drowning the roar of battle at Petersburg. At Bermuda Hundred, upriver, soldiers thought another mine had been exploded in the trenches, and in Richmond, nearly twenty miles away, it was feared that the arsenal had blown up again. Shock waves raced through the air, along the ground, and across the river, knocking over buildings, wrenching ships from their moorings, and creating landslides on the bluffs. Cannon, men, shells, horses, and planking flew through the air. Ships in midstream fled down the river; workmen, laborers, and soldiers panicked, joining the horses and cattle which stampeded through the streets.

A 12-pounder shell tore through the cardplayers' tent, crossed the bed, and landed in a campchest. The men jumped up and dashed outside just as a second shell exploded overhead. Clitz led the van of officers as they scurried for cover. A jagged shell fragment whizzed past Schaff's left shoulder. He pulled up short, reminded of his responsibility, and headed for the wharf.[4]

In Grant's headquarters a terrific shower of bullets, shells, and boards rained through the roof. The general remained seated, nervously chewing a cigar. His aide, Colonel Orville E. Babcock, who was standing next to him, was struck in the right hand by a bullet. Just outside, a mounted orderly was killed outright, three others were wounded, and several horses were slain. Miraculously the commanding general was unharmed and, after learning where the explosion was located, he wrote out a telegram to the War Department: "Every part of the yard used as my headquarters is filled with splinters and fragments of shell. . . . the damage to the wharf must be considerable. . . ."[5]

Mr. Baxter, the lemonade and syrup man, had been surrounded by mule drivers, soldiers, and laborers. A McClellan saddle, from a canal boat loaded with equipment from Sheridan's command, bowled through the crowd, knocked down customers, and struck Baxter in the stomach, killing him instantly. The barber, his tent, his chair, and the "nexts" lined up outside simply disappeared. Schaff's clerk was hit by a shell fragment which sheared off the top of his head. The repatriated Confederate was seen floating in the river the next day, and, a half-mile from the wharf, a musket stood with its bayonet buried in the earth—all that remained of a waterfront sentinel.[6] At the top of the bluff, Morris Schaff surveyed the scene. Later, he could find but one word to describe what he saw: "Staggering."

The bow of a schooner had been blown through a storeshed, and in the river the supply ship *Lewis* burned fiercely. Prostrate figures of men and animals littered the earth; piles of ammunition and guns, once neatly arranged, were scattered like straw and grains of sand cast upon the ground by a giant. The main warehouse, which had been six hundred feet long, was now totally destroyed. In the wreckage Schaff found the body of Sergeant Harris, an old and faithful comrade; nearby, a Corporal Bradley implored Schaff to move the timbers pinning him to the ground.

Suddenly someone shouted, "There it goes again!" and the would-be rescuers ran like the wind—all but Schaff, who spotted the fire licking at a pile of ammunition. He dashed over and frantically beat at the flames with his forage cap.

The boat that exploded first was loaded with several thousand artillery shells; this ignited a hundred thousand rounds of small-arms ammunition in adjacent packing cases. The combined power of these explosions in turn set off other stores of munitions. A survey of the damage was admittedly incomplete, but even the most modest of figures—58 killed, 126 wounded, and property damage in excess of $4,000,000—were catastrophic.[7]

An investigation was conducted according to military principles, but actual witnesses were lacking; everyone who could have known anything about the incident was dead or missing. The officers could only make an "educated guess" as to the cause. It was known that the explosion originated on a barge loaded with artillery ammunition, where the cargo was being stevedored by former slaves. These contrabands did not have a record of being especially careful as they carried shells to the ordnance yard; a strong blow on a shell could have begun the disaster.[8] It was not until after the war that the real origin of the explosion was discovered.

Captain John Maxwell of the Confederate Secret Service was a courageous and inventive operator. In Richmond, he had perfected a small clockwork torpedo which could be activated by a dial and lever to explode at a predetermined time.[9] In company with R. K. Dillard, he left Richmond on July 26 with the "machine" and twelve pounds of powder disguised in a wooden candle box. They traveled overland by night to Isle of Wight County on the other side of City Point, looking for a chance to place the bomb aboard a Union ship. Learning of the immense supplies gathered at the City Point depot, they retraced part of their route and reached the base before dawn of August 9. The dauntless operators crawled through picket lines to the top of a bluff a half-mile from the river, where Dillard remained while Maxwell walked down to the wharf. Maxwell sat down on the bomb, outwardly dozing, but actually watching for an opportunity that would allow him to place his box aboard one of two barges he saw discharging ammunition.

Finally, the captain of a barge came ashore. Maxwell saw his chance, picked up the bomb, and started for the boat. The sentry, a German, spoke little English; to confuse him Maxwell, a born linguist, spouted a volley of Scottish dialect, and an impasse quickly developed. Using gestures, the disguised Confederate succeeded in being passed, and delivered the explosive to a Negro aboard the barge, saying that the captain wanted the box stored below.

In their retreat on the bluff the Rebels were severely shaken by the explosion: Maxwell recovered from the shock, but Dillard was deafened permanently. Regardless, they were successful beyond their wildest expectations. Maxwell's report was that of a proud operator—with one regret: "According to the report of the enemy, a party of ladies was killed. Of course we never intended anything of the kind. . . ." He needn't have reproached himself on that score: with exceptional forethought he had sat on the wharf until a steamer loaded with passengers left for Baltimore; in addition, he had set the mechanism to explode when most of the laborers would be at lunch.[10]

Like many Confederate attempts at discommoding the enemy behind the lines, the venture at City Point lost much of its effect because it was not related to any other event. If it had been timed to coincide with an attack on the Petersburg lines (Lee made such an attempt the following spring), it might have been more than a purely local success. As it was, it was like a boulder dropped into a large river, which produces an immediate effect but does not slow the current.

Maxwell and Dillard were members of the Confederate Secret Service Corps, a group organized by Captain Thomas E. Courtenay, forty-one-year-old Irish cotton broker from St. Louis. Joining the Confederate Army, he served in the West under Generals Sterling Price, and E. Kirby Smith. In August, 1863, Special Orders No. 135, issued by the headquarters of the District of Arkansas at Little Rock, authorized Courtenay to enlist a secret service corps not exceeding twenty men. Just what they were to do is not clear.[11] Courtenay himself seems to have traveled between Richmond and the West, evidently as a courier. In December, 1863, he proposed an entirely separate secret service corps, which he said was "to be employed in doing injury to the enemy" especially on the Trans-Mississippi streams. He also suggested sending agents into the Union, to the West Indies, and to Europe to destroy Federal steam-powered vehicles.[12]

He had worked on a new kind of torpedo: a lump of cast iron, serrated and painted to look like a piece of coal. Inside it was a miniature torpedo—a fuse and an explosive. This deceptive mine would become the principal weapon of his Secret Service Corps. His castings were so realistic, he wrote, "that the most critical eye could not detect it." One especially critical pair of eyes did study them: "The President thinks them perfect." But Courtenay could not use them as soon as he wished, for Secretary of War Seddon refused to organize the new corps until he had the approval of Congress. Courtenay drafted a bill that would permit the Secretary to raise the corps, the members of which were to receive no extra pay but rather a percentage of the value of property they destroyed or damaged. (If he could not get Congress to cooperate, he was prepared to let "a most respectable gambler" buy into the project. He did not especially like the idea, he said, but would take "any port in a storm.") The bill was signed by the President on February 17, 1864; Courtenay was told to raise a unit of twenty-five men. The weapons were to be prepared in the laboratories of the War Department which was to furnish all the chemicals used. In addition to the coal torpedoes, the corps developed an explosive that fitted in sticks of wood, and the clock torpedo designed by John Maxwell. The organization was specifically prohibited from attacking any passenger vessels or railroads or subjects wearing flags of truce.[13]

Apparently, Courtenay never used his torpedoes, despite an order by Seddon telling Major General W. H. C. Whiting at Wilmington, North Carolina, to help the inventor get a barge of coal—laced with the torpedoes—into the hands of the blockaders. The Federals were alerted when, on March 19, a courier was captured by the U.S.S. *Signal* while trying to cross the Red River in Louisiana. Among the incriminating data in his pouch was a letter from Courtenay describing the invention. The pack was taken to Admiral David D. Porter at Alexandria, Louisiana, who issued General Orders No. 184 on March 20, warning of the torpedoes and saying that anyone caught with the weapons would be "shot on the spot." Courtenay and his family, who were living in Maryland, moved to Nova Scotia and then to England to escape possible Northern vengeance.[14]

There were, however, several mysterious explosions which may have been the work of Courtenay's men. One occurred on April 15, 1864, as the *Chenango*, a paddle-wheel gunboat, was leaving New York harbor for Fort Monroe, Virginia. Her port boiler exploded "with terrible violence," killing one crewman and scalding nearly three dozen others, twenty-two of whom died. No satisfactory explanation was forthcoming. The boilers were brand new, having been installed at the Brooklyn Navy Yard, and were of "Martin's Patent," a type widely used by the Navy. There was no previous difficulty with them on record. The first report of the investigating committee was weak indeed, suggesting "a defective vein of iron" as the probable cause. This was changed some months later by another verdict that placed the blame on the use of too much pressure in an imperfectly braced boiler. Eleven years later, with an ocean between him and Federal authorities, an anonymous writer in England (probably a former member of the corps) claimed that one of Courtenay's coal torpedoes had exploded the boiler, but he offered no definite proof—indeed what proof could he offer? Whether the explosion was an accident or intentional will never be known.[15]

After the fall of Richmond, Brigadier General E. H. Ripley,

U.S.A., found a sample of the coal torpedo that had been given to President Davis in the President's "private cabinet." [16] The discovery of these weapons gave rise to closer examination of the records of ships sunk or damaged under unusual circumstances. Several vessels are suspected of having been possible targets. At least one, the *Greyhound*, was destroyed—in the James on November 27, 1864, and the passenger ship *City of New London* was attacked the day before. One former Rebel even hinted that the *Sultana*, which exploded and sank on the Mississippi when loaded with released U.S. prisoners of war in 1865, was also a victim of the coal torpedo, but no substantiation has been found. [17]

The explosion of the *Greyhound* was the most spectacular and, like the City Point disaster, nearly killed several top-ranking Federal officers. Lean and graceful like her namesake, this luxuriously appointed steamer was General Benjamin F. Butler's headquarters ship. On the morning of November 27, she was headed down the James en route from Bermuda Hundred to Hampton Roads. On board were three important passengers: Butler, General Robert C. Schenck, who was "taking a little excursion for . . . his health," and Admiral David Dixon Porter, on his way to meet the Assistant Secretary of the Navy. The *Greyhound* had been under way no more than half an hour when, at Porter's request, Butler had her turned back in order to land several "cutthroat-looking fellows" the admiral had seen in the saloon. After a thorough search for stowaways, she started again.

Some seven miles below Bermuda Hundred the ship was shaken by an explosion. Clouds of heavy black smoke boiled from the engine room, followed by screeching steam as the engineers threw open the safety valve.

"What's that!" exclaimed Butler.

"Torpedo!" shouted Porter, "I know the sound!" Together, they rushed on deck. Flames leaped from amidships and the smoke smelled strongly of coal tar. Forward, the crew was already diving into the muddy water. Porter ran with the ship's steward to the Captain's gig which hung on the port side. He raised the small boat with his shoulder while the steward tugged at the tackle. Together they lowered it as another was launched on the starboard side, and the senior grade officers climbed aboard. The captain ignored the flames to rescue his commander's papers, burning a hand in the process. At a safe distance, the officers watched the ship burn, while near-human screams from Butler's fine horses mingled with the sound of escaping steam and the roar of the fire. Slowly the *Greyhound* settled, sizzling; she came to rest on the bottom, a total loss. The cause was later thought to be a coal torpedo which exploded in one of the furnaces. How it got into the coal bins has never been learned.[18]

On November 26 an unidentified torpedo with its fuse burning was found in the berth of the steamer *City of New London* just before she was to leave her dock in New York City. The operator was never caught.[19] A similar "infernal machine" is reported to have been developed in the North, but supposedly none were used against Confederate blockade runners.

No sample of the coal torpedoes is known to have survived. Yankee threats of severe punishment probably prompted Southerners to destroy any in their possession. Few were the men who admitted using the weapons, and those who did were requested by other Confederates not to divulge their activities. Most of the operators felt fairly safe, for the files of the Secret Service Corps were destroyed at the end of the war during the burning of Richmond.

CHAPTER XIII

STALEMATE ON THE JAMES

Along the upper James a naval stalemate of magnificent proportions had developed—a fact which both sides tacitly admitted. Below Dutch Gap the Federal ships reigned, secured from enemy attack by a mighty system of booms and barriers including piles, rafts, sunken hulks, and mines and guarded by shore batteries of heavy cannon. Monitors were anchored below, and between them and the obstructions, small, swift vessels plied, watching for signs of floating Confederate torpedoes or indications of an attack by the ironclads of the Confederate Navy. Their vigilance was not wasted, for, as both navies knew, there were enough capital ships upriver to wreak untold havoc with the Union squadron if they ever got loose.

Thus, two of the most powerful naval flotillas in the world remained practically immobilized in this narrow stream. The Southern squadron represented the Confederacy's greatest concentration of ships. It was built around three large, casemated, ironclad rams: the *Fredericksburg,* the *Virginia II,* and the *Richmond.* With them were several gunboats of varying value, including the *Beaufort,* a veteran of the *Virginia-Monitor* battle, the *Hampton,* the *Nansemond,* the *Roanoke,* and the *Drewry.* Most of them were lightly armored and mounted few guns. Of more interest, however, were the swift, barbed torpedo boats. They were all forty-six feet long and had appropriate names: *Hornet, Scorpion,* and *Wasp.* In addition there were the older tug *Torpedo* and the *Shrapnel,* sometimes used as a torpedo boat. (The *Squib* had been sent overland to Wilmington.) This armada was commanded by Flag Officer J. K. Mitchell who was pugnacious, realistic, and offensive-minded—but not as argumentative or colorful as his opponent David D. Porter, who had been brought out from the Mississippi.

As had happened so often in warfare, both of these men's thoughts were nearly ·parallel. On October 4, 1864, General Robert E. Lee sent Mitchell an intercepted Union message which said that the Federals were putting percussion torpedoes in the river. The Confederates began to fit to the bows of their warships grapnel hooks similar to the "torpedo rafts" the Yankees had been using. Just a few days later, Admiral Porter, acting on similar intelligence about the Rebels, issued a general order to make certain that all *his* ships were equipped with torpedo rigs too.[1]

Both men were aware that simply sitting in their powerful ships, menacing one another across a double line of obstructions, would in no way contribute to a decision. True, the mere existence of the Confederate James River Squadron was a "fleet in being" and immobilized an equally strong Union force—ships that otherwise would have been employed at Wilmington or Charleston or Mobile. Several times during the past year Mitchell had ventured to duel with the Federals, but now time was on their side, and if he was going to move, he must move soon. An attack offered brilliant possibilities if it succeeded: if the ironclads could get down to City Point, it was conceivable they would be able to destroy the base of the Army of the Potomac and, assisted by Lee from Petersburg, cause Grant to withdraw his army. At the very least the ships might put a Federal squadron out of action and cut the enemy's pontoon bridges across the James. But until the

Rebels were ready the Union ships had to be kept at bay.

In late October, 1864, Mitchell contacted Gabriel Rains—now in charge of all torpedo operations. The ships, he said, would be ready to set torpedoes in the river the next afternoon "and [I] will thank you for a supply, with . . . an adept [operator]." Rains's man worked on the project with Beverly Kennon and his crew. They placed booms of logs chained together across the channel at Bishop's Bluff, but left gaps thirty feet wide at each bank. Mitchell wanted to protect these approaches from Federal row or torpedo boats with small, floating "self-acting torpedoes."

To guard the approaches, he fitted out cutters of his own and equipped them with miniature spar torpedoes containing only five to six pounds of powder. At Mitchell's request they were devised by Commander John M. Brooke at his laboratory in the Navy's Office of Ordnance and Hydrography in Richmond. It was considered likely that these boats would encounter no enemy vessel larger than a similar launch, and as they operated mostly at night, often in fog, it was possible they might collide with some larger Confederate ship. The small torpedoes, while large enough to damage rowboats, would not hurt the larger friendly vessels.[2]

On November 8, Mitchell suspected that his ironclads were in danger of an enemy torpedo attack at their anchorage near Bishop's Bluff, and he moved them upriver for safety, at the same time stationing gunboats to supplement the spar-rowboats. In early December Kennon finally plugged the gaps left in the obstructions because of "recent indication of active operations on the part of the enemy. . . ." He set up two underwater plantations of stationary, frame torpedoes, using the small spar type. They were weighted so that they were about eight feet beneath the surface at high tide. Seventeen were placed below Battery Semmes and twenty near Battery Dantzler across the river. In addition, the Submarine Battery Service established several more electric mine stations with their 2,000-pound iron tanks.[3]

The two commanders were again approaching tactical problems in a similar manner, for at nearly the same time (December 2, 1864) that the Confederates were making these preparations, Admiral Porter issued the following order to Commander W. A. Parker, of the Fifth Division, North Atlantic Blockading Squadron:

The picket boats must always be kept in readiness at night, with their torpedoes ready for service, and if an ironclad should come down they must destroy her, even if they are all sunk. For this . . . you must select men of nerve to command them. . . . The locomotive light must be lit on the bow of the *Onondaga* and the torpedo catcher fitted . . . at all times.

I have an idea that one of these rams can be blown up by two men, with cork [life] jackets on, getting above them or below . . . and floating down . . . with the current. Now, I propose a long, light pole, 30 feet long, with a torpedo at the end, supported at 10 feet from the end by bladders or gutta-percha life-preservers and a life-preserver at the other end. . . .

Though Porter thought it was doubtful that the daring men with the explosives would be killed, they would probably be captured and, for that reason, should be entitled to the full value of any ship they helped sink.[4] The reason for the letter was that Porter was leaving to take part in the first attack on Fort Fisher, North Carolina, and he was taking with him most of the heavier ships of the North Atlantic Blockading Squadron. Parker would have only the monitor *Onondaga,* with her fifteen-inch guns, and light torpedo boats and gunboats. The U.S. entry in the torpedo boat field—the *Spuyten Duyvil,* carrying eighty torpedoes and a pair of clockwork explosives—was on its way to the James. She reached Aiken's Landing on December 15.[5]

A watchful uneasiness was maintained for several weeks while the Confederates prepared for a grand assault using every ship they had. They were ready by January 24, 1865, when a freshet poured down the James, raising the ice-choked waters. During the night the Rebel ships took their positions. The ram *Fredericksburg* led, accompanied by the *Hampton* which had the torpedo boat *Hornet* lashed to her port side. Next came the *Virginia* and the gunboat *Nansemond,* the *Torpedo* with the *Scorpion* lashed to her starboard, followed by the *Richmond,* which had as her consorts the gunboats *Drewry* and *Beaufort,* with the little *Wasp* attached to the latter. There was no room for maneuvering; the attack could only be head on, but if they could crack the barriers and survive the pounding of the shore batteries, they might destroy the Federal ships.

Slowly they steamed in the cold predawn, lights hooded, crews at general quarters beside loaded guns. There was not a sound except the "natural" ones—the hiss of steam, the chopping of paddle wheels, the gurgle of water, and the clanking of well-oiled machinery. A mishap occurred as they neared the obstructions. The *Torpedo* was forced aground by the *Virginia*, but was freed with help from the *Nansemond* and the *Drewry*.

At Trent's Reach the first obstacles were encountered. Lieutenant C. W. Read wrote:

When we arrived near the obstructions, Capt. Mitchell brought the fleet to anchor. He then went in the *Scorpion* with Flag Lieut. Graves and myself. We went down and sounded through the obstructions. . . . While . . . sounding, the Federal picket boat discerned us and gave the alarm. As the enemy occupied both banks, a heavy fire of big guns, field-pieces and muskets, was opened on us, and a perfect rain of missiles swept over our heads. Capt. Mitchell was the coolest man under fire that I ever saw; he stood by the man at the lead, and was not satisfied until soundings had been made many times across the gap in the obstructions.[6]

The great mass of the *Fredericksburg*, piloted by Mitchell, passed through. Mitchell then boarded the *Virginia* and she tried to follow, but ran aground. The same fate befell the *Richmond*. As if by a signal, everything began to go wrong. The *Beaufort* knocked off the *Virginia's* torpedo spar trying to free her. Read, in the *Scorpion*, retrieved the spar and put it back on the bow.

Daylight revealed the Confederates to their enemy who stepped up the cannonade as the *Fredericksburg* rejoined the squadron. Then the *Scorpion* crunched aground and remained fast. Read, in the *Hornet*, tried to pull her off but failed. He then called for the *Wasp*, but she had no better luck. Before anything more could be done, the wooden gunboat *Drewry* suffered a direct hit and exploded with a mighty roar. The nearby *Scorpion* received part of the blast which killed two of her crew and wounded three. Federal batteries dueled with the ships all day, assisted by the *Onondaga*, one of whose 15-inch shells pierced the *Virginia*. At last rising water floated the ships and they pulled out of range.

That night, January 25, 1865, Mitchell was ready to continue his attack and intended to ram the *Onondaga* with one of the torpedo boats, but the pilot of the *Virginia* which was in the lead, refused to take her through the gantlet of Union shells. Mitchell sent Read down to recover the *Scorpion;* however, the Yankees expected just such a move and kept a light shining on her which prevented her being refloated by the Rebels.

Where were the Federal ships during most of the battle? This question was asked by many of those people responsible for the defense of the river. When rumors of the attack reached down-river, a flurry of preparations had begun, including the readying of the U.S. torpedo boats. When the Rebels appeared at the obstructions, telegraph wires hummed with urgent messages. Grant notified Assistant Secretary of the Navy Gustavus Fox who ordered Parker to send his ships to block the river. Nothing happened. Grant signaled anxiously: "He seems helpless," and took control, directing Parker to move. He was backed by Secretary Welles who asked, "Where is the *Spuyten Duyvil* torpedo boat?" Fox, too, wondered: "I do not understand where our torpedo boat is. She ought to dispose of all the rams. . . ." [7]

Belatedly, at daylight on January 25, Parker sent his vessels into battle. The *Onondaga* drew close enough to help the shore batteries, but the torpedo boats never gave any help. The *Spuyten Duyvil* was with the *Onondaga*, but she had no torpedo since the powder was wet from being stored in a cave. Her skipper heard the battle, but did not participate. The powder in the *Duyvil's* torpedoes was replaced, too late to be of any use. This situation became the subject of much recrimination. Admiral Porter had left explicit instructions in the event of such an attack:

> I would like nothing better than to have the rams and rebel gun-boats come down . . . I should certainly expect a report that they had all been destroyed if that torpedo boat is what she professes to be, and not a humbug as I have found all such contrivances. You should be able to whip the whole rebel Navy. Your little torpedo boats should be able to whip a ram apiece, and if my instructions are carried out they will always be in readiness. [8]

He was furious at the report of inactivity. Parker's failure to expose himself to battle and his refusal to attack the enemy ships led to the commander's court-martial. His defense—that he had retreated to get room to maneuver—was not accepted, and he was sentenced to be dismissed from the Navy. Welles disapproved of the verdict because of a technicality and had Parker placed on the retired list. [9]

Once the Confederates had definitely withdrawn up the river, the Unionists set about making a prize of the helpless *Scorpion.* Lieutenant J. W. Simmons of the *Eutaw* sent Ensign Thomas Morgan in a launch with eleven seamen to get her at 7 P.M. on January 27. They rowed past the obstructions and reached the vessel an hour later. She was refloated without difficulty. Her two engines were comparatively sound, and a pair of torpedoes were ready for use. Proudly, Morgan towed her back to the *Eutaw.* Northern engineers were lavish with praise over her builders' "admirable workmanship" and by the end of the month she was serving as a tender for the *Onondaga.* Shortly, because of the difficulty in forcing a passage up the Roanoke River, Porter had her sent to that area to help.[10]

Intelligence reports indicated another Confederate torpedo boat attack. Brigadier General C. K. Graham asked Commodore J. F. Schenck of the *Powhatan* to help him locate the boats, and on February 6, the gunboat *Delaware,* two armed launches, and 150 soldiers in boats set out to explore Jones, Chuckatuck, and Pagan creeks across from Newport News. Lieutenant George W. Wood, U.S.N., landed troops a mile below the mouth of Jones Creek and then went up the waterway with the boats. On the left bank about a mile and a half upstream, he saw a spar torpedo weighing seventy-five pounds, which was all ready for use. At the head of the creek the men searched a house and burned a small sloop. On the way back they learned that a torpedo boat was supposed to be hidden up a small branch of the creek, and, carefully probing the narrow, reed-choked water, they found a schooner's yawl three-quarters of a mile from the James. She was equipped with a long, sturdy oak beam on the bow and was clearly intended for use as a torpedo boat.[11]

On January 30, a similar incident occurred in King's Creek, down the river. A lookout on the little gunboat *Henry Brinker* spotted a sloop there, and two boats were sent to investigate. On their way back, after deciding the boat could be of no use to the enemy, they saw five men running into the brush. The boats gave chase until ice in the creek prevented their going farther, then the sailors climbed out and continued the race ashore. The fugitives escaped, but a pair of suspicious-looking dirt mounds were found. When opened, they revealed two 150-pound mines "with a supposed galvanic battery." (Actually, one of these was a simple friction primer torpedo and the other had three percussion fuses.) [12]

After the near-success in crashing through the obstructions, the Federals reinforced them and took care to post more monitors below them. The Confederates had not been driven back by naval forces, but by a combination of circumstances; Porter was determined that the ships would do the job next time.

Conversely, a similar problem was being discussed by the Rebel sailors. Even though they outnumbered the enemy, they had not been able to accomplish their purpose—hampered mostly by their own ineptness in repeatedly running aground. There was still a way to attack their opponents, however. "By the use of [torpedoes] . . . which in the Confederacy had been developed to a state of efficiency previously unknown to the world, the clutch of the enemy might be shaken off where it bore hardest," reasoned the C.S. Navy's historian. [13]

The plan was ingeniously simple: attack the monitors from their rear in a movement cloaked under the greatest secrecy. The torpedo boats would be taken overland behind the Union Army's left wing at Petersburg, across the Blackwater River, and launched in the James below City Point. There, the men would capture any small enemy ships they encountered, attach torpedoes to them, and move upstream against the monitors. At the same time, the main squadron would crash the obstructions and join the attack.

Lieutenant Read organized the expedition, and was in command of ninety sailors and marines. The boats were put on four wagon wheels and pulled by four mules. The first day, they reached the Confederate right wing and passed through the next morning in three detachments. They were caught in a sleet storm on the third day and were forced to go into early camp near the Blackwater. There, they learned that their scout had betrayed the plan and an ambush awaited them. The entire command retreated a mile or so and hid in a forest while Read went to the river crossing to investigate. The story was true, and the entire group began to make its way back as fast as possible, expecting to be overtaken by Union cavalry at any moment. It was a harrowing retreat, including wading in waist-deep water, but all of them reached their lines. Seventy-five were sent directly to a hospital. [14]

CHAPTER XIV

THE ROANOKE AND THE RAPPAHANNOCK

The capture of Roanoke Island in early 1862 led to Union domination of the sounds of North Carolina. Those broad, shallow, landlocked bodies of salt water that lie between the mainland and the sea were the scene of a flush of victories at the mouths of waterways to the interior, but the Federals contented themselves with clinging to the coastal area, making only occasional forays up the rivers. During the spring of 1864 this situation was rudely changed in northeastern North Carolina, along the narrow, muddy Roanoke River. There the Confederates mustered the strength to retake Plymouth, a dozen miles upriver, with a brilliantly executed land and sea attack, forcing the Yankees down to Batchelor's Bay, where the river joins Albemarle Sound. The "machine" that played a major part in the victory, and the principal object blocking the river to the light craft of the Union fleet, was the clumsy, jerry-built ironclad, C.S.S. *Albemarle*. Constructed in a cornfield near Edwards Ferry farther up the waterway, she had become the focal point of Yankee plans in this area during the summer. Her sinking on October 27, by Lieutenant William B. Cushing, U.S.N., with a spar torpedo, was the most celebrated torpedoeing of the war. Although Cushing had adopted a Confederate weapon and tactic, this exploit was so publicized that sinkings of Union vessels by Confederate torpedoes in the same stretch of water have been all but forgotten. The destruction of the *Albermarle* did not open the river as the Federals had hoped, however. If anything, the door was locked even tighter, though moved inland a few miles.

Hoping to draw Confederate attention and forces away from Petersburg, in December, 1864, Grant suggested a joint Army-Navy attack upon Fort Branch at Rainbow Bluff, a hundred-foot cliff on the right bank of the Roanoke, some twenty miles above Plymouth. Commander W. H. Macomb, whose Division of the Sounds of North Carolina was a part of Rear Admiral David Dixon Porter's North Atlantic Blockading Squadron, was told on December 1 to cooperate with a thirteen-hundred-man land force in the venture. The Army contributed the Twenty-seventh Massachusetts and Ninth New Jersey Volunteer Regiments, the Third New York Artillery, and a loyal North Carolina cavalry unit. The naval squadron consisted of "double-ended" gunboats—ships like ferries, which could be reversed without turning around—and other light river craft. They were the gunboats *Wyalusing, Otsego, Valley City*, the tugs *Belle* and *Bazely*, and the steam launch *Picket Boat No. 5*, similar to the one used by Cushing against the *Albemarle*. They hoisted anchor at Plymouth at 5 A.M. on December 9, 1864. The coffee-colored river was so narrow that the vessels were forced to proceed in a single column between heavily forested banks. At times red clay bluffs towered above the decks, while at other spots, swampy vegetation grew out over the water's edge. Macomb was assured that the river was cleared and no torpedoes were in the twelve-mile section between Plymouth and Jamesville. As an added precaution, nets had been slung across the bows to scoop up any stray mines. Midway, the engine of the *Valley City* balked, and she exchanged places with the *Otsego*, which moved into second place behind the *Wyalusing*.

Approaching the tiny hamlet of Jamesville about 9 P.M., the ships were signaled to anchor, and, just as they prepared to do so, a Singer torpedo detonated, showering water over the port deck of the 974-ton *Otsego*. A second explosion erupted under her forward pivot gun—a 100-pounder Parrott rifle. It blasted through her decks, throwing the two-ton cannon over on its side. The ship settled sluggishly to the bottom amidst clouds of smoke and steam, coming to rest with her spar deck three feet below the water. Fortunately her crew suffered but a single injury.

The next morning, other ships began to "sweep" the water around the wreck, finding "a perfect nest of torpedoes." The little 55-ton *Bazely* was ordered to Plymouth and on the way stopped by the injured *Otsego* to transfer several men. Suddenly, she too was struck—"blown literally to pieces" wrote an eyewitness. Her captain, pilot, and paymaster were in the pilot house, which was lifted thirty feet in the air, but miraculously, none of them suffered even the slightest injury, though two sailors went down with the ship. Shaken and ashen-faced, her skipper reported to Macomb, telling him that "the *Bazely* is gone up." A member of the commander's staff thought differently: "By that time she had gone down!" [1]

Utilizing the wreck of the *Otsego* for all it was worth, Macomb stripped it of its ten large guns and placed a pair of 24-pounders and two 20-pounder rifled cannon on the hurricane deck, behind cotton bale breastworks, to protect it from Confederate raiding parties. It became, in effect, a tiny fortress guarding this stretch of river, while smaller boats busily dragged the stream. Six mines were dredged from beside the wrecks and a pair were found in the *Otsego*'s net—all in perfect working condition; two of them exploded while they were being pulled ashore. [2]

Having secured the lower reaches of the Roanoke, Macomb sent word for the U.S.S. *Chicopee* to join the squadron, then proceeded upriver, clearing the stream with six rowboats paddling slowly ahead, about 20 feet apart, a chain suspended between them. The *Valley City* and other ships followed. It was "extremely tedious" work, and progress was measured in yards, rather than miles. Caution paid off when a number of explosives were discovered anchored eight miles above Jamesville. On December 14, they discovered twenty-one mines at Shad Island Bend "in the richest and choicest clusters, in some places eight or nine . . . across the river in a line. . . ." These weapons had been placed under the direction of Lieutenant Francis L. Hoge, C.S.N., a daredevil former Union midshipman who had taken part in the captures of the *Satellite* on the Rappahannock and the *Underwriter* on the Neuse River, among other exploits.[3]

It was at this juncture that the land attack faltered. Colonel Jones Frankle, Second Massachusetts Volunteer Heavy Artillery, and his troops had moved ahead along the south bank and probed near the vicinity of Rainbow Bluff, but quickly withdrew to Jamesville and then back to Plymouth without awaiting the arrival of the Navy. He gave as his excuses sickness and a lack of equipment and supplies, claiming that some of his soldiers were actually barefooted in the cold winter weather. He was not abandoning the attack, but would rest and replenish a few days before resuming it, he said.[4]

Then, while dragging the river near Popular Point, a few miles below Rainbow Bluff, the men in the open boats were fired upon by sharpshooters from the shore. Ashore, the marine guard was taken under fire and fell back slowly under the cover of the ship's guns. At the same time, Confederate cannon on Popular Point took the ships under fire. The *Valley City*, in the van, could not bring her guns to bear on the enemy. "Amid a shower of musket balls," her executive officer rowed a line ashore, passed it about some trees, and sprung the vessel's broadside around. She replied to the enemy guns which were located on a hillside some seven hundred yards off. Both the *Wyalusing* and *Chicopee* came up to help the *Valley City*. The former pilot of the ill-fated *Otsego*, who was aboard the *Valley City*, was killed instantly by a musket ball in the head and the ship itself was struck by three shells. One smashed through her coal bunker and lodged within inches of the boiler. Another hit an awning stanchion, and a third grazed the side. Aboard the *Wyalusing* a shell exploded in the wardroom, wounding one man. Prudently, the ships dropped out of range as darkness settled.[5]

During the night Confederate sharpshooters annoyed the Yankees and they got little sleep. Shooting stopped at dawn and all was quiet. Commander Macomb dispatched the *Valley City* to see if the Rebels had moved out. Cautiously, the gunboat approached the bend in the river which was dominated by a high, forested ridge where the Confederates had established their battery. Her bow had hardly cleared the point when she was shaken by three hits in succession from enemy guns in new emplacements barely a hundred yards off. The first shell passed through the pilot house and exploded, killing two men and wounding two more. Two others hit elsewhere, and a fourth whistled past while the paddle wheels chopped the water into white foam as the ship sought to back under the protection of the bend. It was plain the *Valley City* had sailed into a trap, for the Confederates, under Colonel Collett Leventhorpe, had moved their guns during the night. They had determined the range so well that one Union officer indignantly reported seeing a Rebel walk up and pull the lanyard of a gun without even sighting it. The vessels fired into the forest with their 100-pounder guns, but finally gave up and fell back out of range.

On the twenty-second, the *Chicopee* took a shot at the ridge, but received no reply. Then, as she began to go downriver with the others, the Confederates unmasked a hidden battery of 32-pounders and began peppering the ships again. At the same time, riflemen showed themselves and took up positions behind an embankment. Showering the shore with grape, the fleet fell back to the next "reach," running a four-mile gantlet of sharpshooters.

The Confederates were quick to recognize the affair for what it was—the decisive engagement for control of the Roanoke. Colonel Leventhorpe jubilantly reported the success to his headquarters, and no lesser personage than Robert E. Lee telegraphed news of the victory to the Secretary of War. Both Colonel Frankle and Commander Macomb discussed renewing the joint attack, but Admiral Porter told them to call it off. In his report, Macomb told of his difficulties: "As long as the enemy held the banks . . . I would be unable to drag in open boats for torpedoes . . . and could not advance with the gunboats. Had not the torpedoes prevented our advance we could have run past the batteries."

On Christmas Day the Federals destroyed what was left of the *Otsego* by the simple expedient of exploding a pair of captured torpedoes inside her. Total casualties for the futile mission were high: two ships sunk, six men killed and nine wounded.[6]

Macomb's only recourse was to block the river at Plymouth and wait until higher authority decided whether to try and force the river again. He sent a torpedo to Admiral Porter and asked him to have a hundred like it made for use in the Roanoke in case the iron sister of the *Albemarle,* which was being built upriver, should attack his squadron. Porter forwarded the request with several endorsements, specifying electric torpedoes instead of mechanical ones. "I think if we used torpedoes as often as the rebels do," he wrote, "we would soon destroy all their rams."

In the meantime, he instructed Macomb to keep his picket launches operating and equipped with torpedoes, and to use his calcium light on the water at night. The former Confederate torpedo boat *Scorpion* was ordered to the sounds. "Don't let it fall into rebel hands again. It will be a good thing to use their own designs against them," Porter cautioned. If the ram attacked, Macomb was to use his torpedo boats aggressively. "It is easily done," the Admiral explained. "It only wants coolness and decision . . . don't have another disgraceful affair as . . . at Dutch Gap. . . ." However, the *Scorpion* never reached her destination, for on February 13, while under tow from the *Phlox,* she was rammed and sunk by the schooner *Samuel Rotan* near Jamestown Island, taking a crewman down with her. Efforts by the *Cactus* to salvage her were not successful. When he heard of her loss, Porter broadened his definition of a torpedo boat: he wanted Macomb to put spars on "every vessel you have. . . . No matter how many of your vessels get sunk, one or the other of them will sink the ram if the torpedo is coolly exploded." [7]

The stalemate continued until April, 1865. On the eighth, the captain of the *Mattabesett* heard that the new Rebel ram was coming down the river. He dispatched a launch to investigate, but it reported nothing in the river as far up as Jamesville. A boat from the *Iosco* had better luck the next day and discovered the ram wrecked on a sandbar eight miles above Plymouth. The first boat had passed right by it, thinking it was the *Otsego* wreck. The Confederates had abandoned not only the ram but the entire river—probably because of an advance from the north by Union troops. The ram was set adrift on the night of April 5, 1865, and struck a mine. The Federals found her with only a couple of feet of her casemate visible. [8]

From the beginning of the war a small Union squadron, technically a part of the Potomac Flotilla, patrolled the Rappahannock

River some three hundred miles north of the Roanoke, in northern Virginia. Confederate boats constantly crossed to the "Northern Neck," that historic peninsula of the Carters, the Washingtons, and the Lees, ferrying contraband goods and military units or spies, to harass U.S. shipping in the Potomac or at the approaches to the Union capital. Several times the Rebels boldly boarded Federal vessels, but their greatest achievement came in August, 1863, when they captured a pair of U.S. Navy gunboats, the 217-ton paddle-wheel *Satellite* of two guns, and the little 90-ton screw ship *Reliance,* and took them upriver where they became the nucleus of a small fleet located less than a hundred miles from Washington.

Unable to challenge Union strength ship-for-ship, the Confederates resorted to torpedoes. (These were the same waters in which R. O. Crowley had established an electric mine station in 1862.) Rumors of their plans "from the best authority" reached Commodore Foxhall Parker, U.S.N., in April, 1864. (It is of interest to note that this Parker, who was about to be harrassed by torpedoes, was the brother of Captain William H. Parker, C.S.N., who was in charge of the row torpedo boats in Charleston.) It was said that mines were planted at intervals along the southern shore from Bowler's Rocks in Glebe Landing Bay, between Urbanna and Tappahannock, the flotilla's anchorage, all the way to Fredericksburg—some sixty miles all told. One was reportedly exploded by a trafficking canoe and "its destructive powers are said to have been brilliantly illustrated." [9]

In early May of 1864, during the Battle of Spotsylvania Court House, the Army of the Potomac sent its wounded to Fredericksburg and requested that the Navy evacuate some of them down the river. Instructions to send up "a couple of light drafts" and to "be very careful about torpedoes, and send a prudent man to command," were telegraphed to Parker. He immediately adopted a now-common defense—the attaching of wood rakes across the bows of several small ships. Parker wanted to lead the expedition himself, but the Navy Department refused to permit it. The gunboats were retained in the Potomac when it was learned that their draft would allow them to go no closer than four miles from Fredericksburg. The *Freeborn,* a three-gun paddlewheeler, broke her piston and had to be towed all the way back to Parker's base in

St. Mary's River, Maryland. Lieutenant Edward Hooker was sent ahead up the Rappahannock to examine it and report its condition and the locations of any torpedoes.

Hooker's departure was easily meshed into the solution of a problem raised by a report that the Confederates were about to float more torpedoes in Butler Hole (a day rendezvous of the Federals) that very night—May 11, 1864. Parker believed he knew who intended to do the work. "The Peninsula Between the Rappahannock and the Piankatank Rivers is a guerilla haunt which I would effectively break up had I one hundred Marines," he said. In their place, he asked for soldiers.[10]

A young colonel, Alonzo G. Draper, who commanded the military district of St. Mary's from Point Lookout, Maryland, knew Parker well and readily assented to help him. It was not the first such expedition in these waters, nor would it be the last, for the services were constantly conducting small amphibious operations there and probably became as proficient at this kind of warfare as any American units prior to World War II. Draper led three hundred soldiers of the Thirty-sixth U.S. Colored Infantry and fifteen cavalrymen himself. They were loaded on the transport *Star* which was convoyed by the gunboats *Yankee*, *Currituck*, and *Fuschia*. The ships arrived off the south side of the Rappahannock at 7 A.M. on May 12, and, accompanied by a howitzer and thirty-five sailors, the Federals landed unopposed. At the same time, Hooker led launches from the gunboats along the bank and into creeks and coves, dragging for torpedoes.

Hooker's search was successful. He discovered the lanyards to a pair of torpedoes, the operators having fled at his approach. He stationed a boat at each and joined Draper's force which had burned a grist mill. Pickets probing a heavy woods made a valuable find—a torpedo factory, complete with four mines (two of them assembled), powder, tar, beeswax, and other materials necessary to make them. A little later the soldiers caught up with the Confederates and a brisk twenty-minute engagement resulted in the rout of the Rebels, eleven of whom were killed and ten captured.[11]

These mines were of the friction primer type. They were tin cylinders shaped like tubs, filled with fine-grained powder. Inside, at the center, was a friction primer to which was attached a lanyard that led to the shore. Designed to float on their sides, they were buoyed by logs and held at their locations by iron anchors. One officer was highly impressed with what he saw, and after exploding one, wrote that "the appearance was grand, and if a ship was directly over one . . . she would, in all probability be sunk. . . ." He added further, "I feel competent to use the remaining torpedoes against the rebels whenever it is required of me" [12]

Colonel Draper was highly pleased with the success of the expedition and wrote poetically of the cheerful cooperation he had received from the sailors who "denying themselves needful rest and sleep . . . labored night and day . . . aiding the Army in every possible way." [13]

Emboldened by the success of the landing, Parker sent Hooker to Fredericksburg. Hooker telegraphed somewhat cryptically: "Seven feet of water. More would be troublesome less would be better." It was enough, and Parker sent the *Fuschia,* the *Freeborn* and the *Jacob Bell* to him. "Hooker," he said, "is a very careful officer, but the service is a very hazardous one. . . ." The ships proceeded cautiously, launches sweeping ahead, landing parties beating the bushes ashore. In one place they found a Negro who claimed he knew where a cluster of torpedoes was located and guided them through the area. He need not have bothered, for slaves said later they had already been removed. Hooker reached his goal safely at 2 P.M. on May 19, having taken three days to make the journey. The only damage suffered by the ships came when the *Yankee*'s starboard wheel struck a rock. "So impressed were the rebels," wrote Parker, "that . . . they immediately . . . exploded or removed all the torpedoes . . . in the river, above Bowlers Rocks."

Another dividend was the scuttling of the *Satellite* by the Confederates. The Federals found the smoldering wreck and, to complete the destruction, exploded a captured torpedo in the hull.[14]

Yankee domination of the Rappahannock did not last long, for the Army evacuated Port Royal on June 1, forcing the Navy to retreat to its anchorage at Bowler's. On their heels came the Confederates, and it was not long before torpedoes made their appearance once more. To command these operations, the Rebels sent Beverly Kennon, the man who floated the war's first mine near this very area. His reputation preceded him and when Parker learned of the appointment, in November, he requested that Secretary Welles send two more "light draft" ships to use in the river. He spelled out his troubles:

> I feel that I can insure the safety of this flotilla, and especially of the guard vessel off St. George's Island, the valuable store vessel . . . and the large fleet of transports frequently anchored at Point Lookout, from torpedo-boats only by constantly scouring the creeks leading from the Piankatank, Rappahannock, Great Wicomico, Little Wicomico, Coan and Yeocomico rivers, which, with the vessels now at my disposal, I am not able to do.
> I had formed the intention of attempting this with howitzer-launches, but upon a reconnaisance found it to be impracticable, the banks of the creeks being lined with thickets, from which the enemy's riflemen would pick off our oarsmen with impunity.[15]

In New York, a pair of 30-foot steam launches had been fitted with spar torpedoes by Lieutenant William B. Cushing, who intended to use them in an attack on the Confederate ram *Albemarle,* down in the Roanoke River in North Carolina. They left New York on September 22 for Albemarle Sound via the Inland Waterway and, after a series of breakdowns, reached Point Lookout on October 6. Ensign Andrew Stockholm, in charge of *Picket Boat No. 2,* took shelter in the mouth of Reason Creek, in Great Wicomico Bay, thinking it was the Patuxent River between the Potomac and Rappahannock, after his engine had become balky in the face of an unfavorable wind. While repairs were in progress, she was attacked, and Stockholm, trying to escape, ran aground. A short, sharp fight lasted until the sailors ran out of ammunition. The boat and its crew of eleven were captured. The craft was destroyed by the Rebels, who took its 12-pounder howitzer with them.[16]

Meanwhile Kennon was active. On the morning of January 4, 1865, a landing party on the right bank of the Rappahannock, some six miles from its mouth, uncovered a pair of unloaded torpedoes and two barrels of powder. Evidently the Federals had surprised Kennon's men in the act of floating the mines, but most important, Parker learned of "some deviltry in the Rappahannock" that sent a wave of anxiety through the command.

The "deviltry" was thought to be a Confederate attack upon the triple-turreted ironclad *Roanoke* at Point Lookout by torpedo boats carried overland and launched in the Potomac. Secretary Welles cautioned her skipper Captain A. H. Kilty: "It behooves you to neglect no preparation or plan for preventing the insurgents from accomplishing their nefarious object." Kilty lost no time. He rounded up timber, cut it into spars, and ringed the ship so that they supported heavy nets, enclosing her within a twenty-foot circle. The *Roanoke*'s cannon were manned around the clock while Parker patrolled the southern shore of the Potomac with small craft. On January 9, the *Banshee*, a sidewheeler mounting five guns, was added to the flotilla and, on the following day, Kilty asked for three more 12-pounder howitzers to use in repelling torpedo boats.

No attack developed; none had even been intended. The whole thing was based on the report of a Confederate deserter who saw preparations for the James River attack. By March, 1865, the Federals, by careful maneuvering, had cleared the Rappahannock once more and, as events proved, for the last time. No more was the torpedo a menace in the waters of northern Virginia.[17]

CHAPTER XV

"DAMN THE TORPEDOES!"

During the early part of the war, Farragut's West Gulf Blockading Squadron had worked the waters around the mouth of the Mississippi, captured New Orleans, and forced its way up the river past the powerful batteries at Port Hudson and Vicksburg; but the admiral had long chafed at the Union high command's refusal to allow him to attack and capture the fine harbor at Mobile, Alabama. Here, in a sheltered bay some thirty miles long, was one of the Confederacy's greater cities, and in the harbor were a variety of blockade runners and warships of the Confederate Navy. Finally, in 1864, Farragut was sent instructions to attack Mobile.

Conditions there were similar to those at Charleston: Farragut's fleet cruised outside the harbor, barred from entering by strong fortifications and lines of torpedoes. Inside was a Confederate squadron built around the powerful ironclad ram C.S.S. *Tennessee*, under the command of the Confederacy's only full admiral,

Franklin Buchanan, who had been skipper of the C.S.S. *Virginia* in her first battle at Hampton Roads. Farragut was determined to force his way into the harbor, regardless of cannon, ironclads, or the torpedo defenses planted by Rains that February. He had run the batteries at Port Hudson and Vicksburg, and felt he would be able to force his way through this pass as well. The Confederates were equally determined to prevent him from doing so.

Farragut was fully aware of the great risk he was running, and he also knew about the torpedoes that lay in his path. Through his spyglass he could discern the ramparts of Fort Gaines on Dauphin Island at his left, and Fort Morgan at Mobile Point to his right, both bristling with tiers of artillery in their embrasures. There was also a dark line of piles stretching east from Fort Gaines to the ship channel, and he knew that scores of floating mines, invisible from the surface, were anchored in the wide channel. The Confederates had placed about 180 of the tin Fretwell-Singer models and Rains keg torpedoes there, and on the bottom rested several great electric tanks full of powder. Many of the smaller explosives were in the main channel; the remainder were staked out a little farther west—but between Fort Morgan and the eastern limit of the mine plantations there were none. This gap had been left open to permit the entry of Confederate blockade runners, but it was safely under the range of the guns of the fort.

Here was the key that the Union commander hoped would open the Confederate door: he intended to send his ships through this gap, avoid the clusters of torpedoes, and take his chances with the cannon. He was concerned about the torpedo menace, and for several nights, cutters under the command of Flag Lieutenant John Crittenden Watson had rowed out to examine the lines of explosives. Evidently, they actually succeeded in removing and cutting some of them adrift. However, there were three lines of explosives, and it is doubtful whether their efforts were of any great benefit. One fact they may have discovered was that many of the Singer torpedoes were ineffective due to overlong submersion. Early on the bright, pleasant morning of August 5, 1864, Farragut's fleet moved to the attack in parallel columns—one composed of fourteen wooden vessels lashed together for mutual protection, the other made up of the ironclad monitors led by the *Tecumseh*, 1,034 tons with two 15-inch smoothbore guns in her

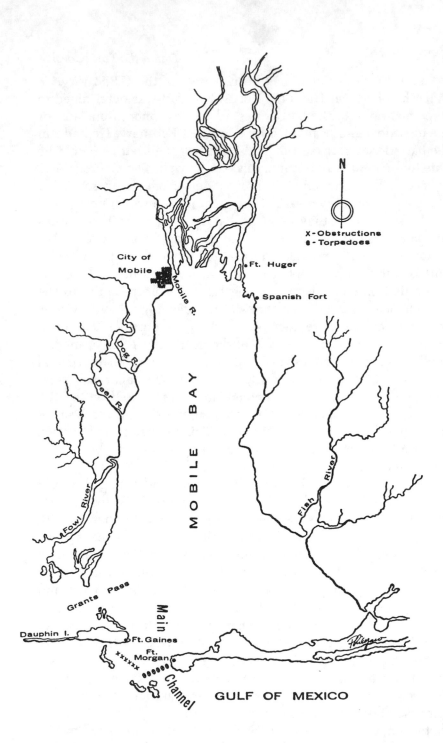

N

X – Obstructions
● – Torpedoes

City of
Mobile

● Ft. Huger

● Spanish Fort

Mobile R.

Dog R.

Deer R.

Fowl River

M O B I L E   B A Y

Fish River

Grants Pass

Main

Dauphin I.

Ft. Gaines

Ft. Morgan

Channel

GULF OF MEXICO

MOBILE BAY

turret. She was skippered by Commander T. A. M. Craven, a capable, experienced officer; no finer ship or officer could have been found to start the battle.

The attack began at precisely 7:07 A.M. when the ships drew within range of Fort Morgan. The *Brooklyn,* leading the wooden column and mounting two dozen guns, replied to the Confederate fire, and the shooting became general. Suddenly, at 7:30, the ironclad *Tecumseh* veered to the left, causing the whole line to stagger, and in a matter of seconds sank completely out of sight. Before the *Manhattan,* which was following, could stop, she passed right over the wreckage and survivors of the sunken ship. During the rescue, the Confederates held their fire.

Ashore, the Confederates were as thunderstruck as the Union sailors to see this great ship vanish. One moment she had been "pushing on gallantly," then she simply raised from the water and nosed under, her propeller thrashing as her stern slid beneath the waves.

A torpedo had exploded directly under the monitor's turret, blowing out part of her bottom. Witnesses claimed she disappeared in half a minute. When a diver explored the wreckage later, he found the bodies of many of her crew still at their stations. Only 21 of her 141 officers and men escaped. Without question, this was the most completely disabling blow struck by torpedoes during the entire war.

Clinging to the rigging of the *Hartford,* Farragut viewed the disaster—as grim and determined as before. Then the *Brooklyn* began to back, confusing the line. She reported that her maneuver was due to the underwater explosives that lay ahead.

"Damn the torpedoes!" shouted the admiral. "Four bells. Captain, go ahead!" The ships' bottoms actually brushed across some of the mines. Inside the vessels, frightened sailors heard the steel rods of Singer torpedoes snapping against the primers as the Yankee ships triggered the springs. Miraculously none exploded, and the fleet broke through the barrier to engage and defeat Buchanan's ships, capturing the battered *Tennessee.* The admiral himself reportedly heard the torpedoes snapping above the din of battle. No one knows why the mines did not explode, but evidence points to damp powder and frozen trigger springs as the cause.[1]

The Yankees' forcing the harbor entrance did not hinder Confederate torpedo operations at Mobile—if anything, it was responsible for increasing them. All commands busily planted more barricades at the approaches to fortified places (in water as well as on land) with the realization that an enemy attack was a possibility at any time. On the day after the battle a load of twelve keg torpedoes was sent from Mobile to be placed in the mouth of Dog River not far from the city. In the charge of Lieutenant F. S. Barrett and two Negroes, they were carried in a four-mule wagon. One of the Negroes drove, while the other, riding in the rear, "amused himself" by unscrewing the safety guards of the machines. Suddenly the lieutenant, who was riding a few paces ahead, heard several explosions in rapid succession and found himself blown from his horse into a briar hedge that lined the road. Looking up, he found the wagon blown to bits, "three mules dead in the road . . . his horse running wildly about, the . . . driver dead . . . and the author of the mischief a dozen steps off on the beach, mortally wounded." [2]

Despite the accident, Barrett succeeded in blocking the Dog River within three days, and the Blakely and Apalachee a few days later. Other mines—as many as fifty-four at a time including keg, Singer, and frame torpedoes—were put in front of the batteries opposite Mobile, such as Battery Huger or Battery McIntosh. At the same time, the Yankees were engaged in removing torpedoes from the bay (five men were killed and eleven wounded in the process). [3]

In December, one of these weapons claimed a victim. The U.S.S. *Narcissus*, a tug of 101 tons doubling as a light gunboat, was ordered to proceed near obstructions off Mobile for picket service during the night of December 7. Her commander, Ensign William G. Jones, followed instructions and anchored in eight feet of water at 10:30 P.M. A sudden squall that blew in from the northeast punished the ship considerably—so much so that Jones feared she would be blown aground. He got her under way and moved out about a mile, anchoring in deeper water. "While paying out chain," it was reported, "the vessel struck a torpedo which exploded, lifting her nearly out of the water and breaking out a large hole in the starboard side, amidships, besides doing other damage." The ship was filled with steam which poured from a burst pipe and drove men from below. Fearing the boilers would explode, Jones ordered Second Class Fireman James Kelly to "haul the fires." This Kelly did at great personal risk, wading through hot water which scalded him badly. An officer and two other men were scalded as well.[4]

CHAPTER XVI

"THE TORPEDO IS JUSTIFIABLE"

The published accounts of torpedoes & mines in Mobile Bay, Florida rivers & seacoast," wrote an enchanted Rebel, "read like fairy tales, and many a youth was fired enough to make him willing to join any party going on any adventurous trip, without half counting the cost. . . . Setting out torpedoes of all kinds kept many satisfied and hard at work . . . constructing and placing them. . . ." [1] By late 1864 the spectacular success that torpedomen had scored against the enemy was widely known, and even in Europe naval engineers were beginning to study these new weapons seriously and to set up torpedo bureaus of their own.

Land torpedoes were being used almost as widely as water mines, but sinking ships was far more glamorous than the killing or wounding of a few foot soldiers. Consequently, the newspapers did not treat these exploits with the same weight. The Federals tried to use a few of the land mines themselves. Near Savannah, for example, they planned on undermining a railroad bridge with them, and Company I, Third U.S. Colored Troops, was selected to put them in place. Five torpedoes were prepared and taken on the expedition, but it did not come off as expected, and none were used.

It remained for the Confederates to show the Federals how, and the Rebels were directed by the person with more experience than anyone in the world—Gabriel Rains. On June 8, 1864, he was assigned to the "superintending of all duties of torpedoes" and at the same time came into control of some $350,000 that Congress had appropriated for the mines that spring—a far cry from the paltry $20,000 given to the Torpedo Bureau the year before.

When General William T. Sherman ran into Rains's mines in upper Mississippi, he was at first outraged, but by the summer of 1864 he had become resigned—to a point:

I now decide that the use of the torpedo is justifiable in war in advance of an army. But after the adversary has gained the country by fair warlike means . . . the case entirely changes. The use of torpedoes in blowing up our cars and the road after they are in our possession, is simply malicious. It cannot alter the great problem, but simply makes trouble. Now, if torpedoes are found in the possession of an enemy to our rear, you may cause them to be put on the ground and tested by wagon-loads of prisoners, or . . . citizens implicated in their use . . . if a torpedo is suspected on . . . the [rail]road, order the point to be tested by a car-load of prisoners, or citizens . . . drawn by a long rope. Of course an enemy cannot complain of his own traps.[2]

The farther Sherman advanced, the more torpedoes his army encountered. In November President Davis desperately ordered General Howell Cobb to get "every man who can render any service" to obstruct the roads in front of Sherman's army. Colonel George W. Rains at Augusta was to furnish all the specially prepared shells Cobb needed. At the same time, Gabriel Rains was instructed to mine the approaches to Augusta. James H. Tomb of *David* fame was sent down to help as General Joseph Wheeler put torpedoes in the roads leading to Savannah. Routes to Pocotaglio, South Carolina, were similarly blocked. Bridges were destroyed, trees were felled, and subterra shells were put in the center and shoulder of roadbeds, but the mining did not stop the inexorable Union advance.[3]

And then there was the deserter on James Island near Charleston with a story no Yankee dared to verify. With a sweeping gesture he indicated a road in front of their position: "Well, sir, there are torpedoes in that and in a hole dug in the ground there is a can of liquid or Greek fire, with a friction match so arranged that if you step there you will ignite it." [4]

One of the largest concentrations of land mines was found at Fort McAllister, near Savannah. As the Union Army neared that powerful bastion in December, the Confederates put "a considerable number" of land mines in front of the fort and along the causeways leading to it. Most of them were 7- or 8-inch shells buried just below the surface, about three feet apart. When Sherman's forces attacked on December 13, a picket captured on a causeway to the west told the Federals about the buried explosives, but this information could not be sent out to the entire army in time to prevent casualties. Some of the mines exploded as soldiers stepped on them, throwing men in the air, killing twelve, and wounding about eighty. After the battle, Sherman put some of the prisoners to work digging up the unexploded torpedoes. He used the fort's engineers, who probably planted the mines in the first place, and a detail of sixteen men. The Confederate commander, Major George W. Anderson, was infuriated: "This hazardous duty," he said, was "an unwarrantable and improper treatment" of his men. [5] Among the booty at Savannah was one uncompleted torpedo boat still on the ways. The U.S. Navy exhibited considerable interest in her, had her finished and sailed to Port Royal where she saw service as a harbor vessel. [6]

During the last year of the war the Charleston Torpedo Service achieved full-blown maturity. Two years of testing, planning, and attacks had honed its actions to fine effectiveness. The establishment had grown until it required the full-time efforts of from twenty to sixty men. Captain M. Martin Gray was still in charge under Rains's supervision and had his factory at the foot of Hazel Street on the Cooper River. Here copper spar tanks and iron castings for frame torpedoes were made. Electric boiler mines were constructed on the wharf. Fuses were obtained from Savannah and, later, from other sources. These operations were transferred to the Ashley River in 1864 because of Union shelling of the city.

A shake-up took place in August. Gray had "become suspect" for embezzling. It seems he had speculated by buying rope cheaply and selling it to the government for a handsome profit. Evidently, the manipulation came to light when the parties who sold him the rope turned him in for not paying them. Gray insisted (to the Yankees after the capture of Charleston) that his arrest had come from his attempts at being a true friend to the United States. He claimed he had put out empty torpedoes, but his men did not verify the story, nor do the records of his activities. Arrested and court-martialed, he was imprisoned for nearly six months until released by the Federals. Captain E. Pliny Bryan "a particular friend of General Beauregard" replaced him and was assisted by Captain J. A. Simons and a companion.

Many barrel torpedoes were floated during the cold nights of December and January, but an especially important cluster was put out the night of January 15, 1865, on the strength of rumors of an impending enemy attack. Two of the operators, Francis Wood, a native of Charleston who knew the channel well, and Robert Thompson, from Norfolk, were sent out with a boatload to supplement the rope obstructions. Sixteen mines were placed near the barrier and several more at the entrance to Hog Island Channel on the north side of the harbor.[7]

The strongest torpedo boat unit in the Confederacy was also at Charleston. Built and building were nine ships. Unlike the James River group, these vessels were apparently given numbers instead of names. Occasionally they made appearances under cover of darkness, seeking whatever targets they might find; their commanders had rather broad orders authorizing them "to attack the enemy's fleet at any time. . . ." Lieutenant General W. J. Hardee, who was in charge at Charleston, soon realized the potential of the boats and sought to unify their direction by consolidating them into a single unit. "In organizing the corps," he wrote, "I propose to select one officer to control the whole service, who should make his headquarters in this city; one to command a flotilla on the coast . . . and one to command the Georgia fleet. A number of officers are anxious to serve if the organization is perfected. . . ." Among the men were such familiar figures as Commander Isaac Brown, Commander W. T. Glassel, Lieutenants J. W. Alexander and Charles W. Read, and a dozen more. "Unless some action is taken soon by our Government," Hardee concluded, "I fear the enemy will beat us with our own weapons." [8]

As it had several times earlier, the Confederate assumption that a Union naval demonstration was in the offing proved to be correct. The plan was designed to cooperate with the advancing army of Sherman. Admiral Dahlgren's confidential instructions issued January 15 required each ship to lash pairs of fifty-foot pine logs to form an X across its bow and to suspend nets from them. The nets were designed to scoop up floating torpedoes—damage from other types would have to be risked.

A pair of "picket" monitors, the *Patapsco* and the *Lehigh,* moved toward Charleston ahead of the fleet on the evening of the fifteenth, covering a convoy of tugs and other small craft that were dragging with grapnels. Two of the tugs kept close to the *Patapsco,* which led while the *Lehigh* stayed some distance to the side. As 8 P.M. approached, Lieutenant Commander S. P. Quackenbush of the *Patapsco* decided to heave to for the night. He cut the engines when the ship neared the so-called "Lehigh buoy" which marked a shoal where that ironclad had once run aground, and prepared to anchor. However, the monitor drifted with the tide and forced him to call for power again. At 8:10 the hull of the *Patapsco* shivered from a tremendous explosion on the port side near her great turret. The deck lifted and smoke poured from around the edges. Floods of water spewed through a rent in her skin, filling compartment after compartment. Quackenbush, who was standing atop the turret when the torpedo struck, passed word to start the pumps and man the boats, but before the crew could comply, the *Patapsco's* deck slid beneath the water. One boat, the gig with two men aboard, floated free. The monitor hovered momentarily before plunging out of sight; she sank in half a minute, carrying sixty-two officers and men with her.

Some of those who were saved had frightful experiences. In the wardroom, three officers sitting around a table were blown against the overhead but recovered in time to scramble to safety. A man in the windlass room saw a flash and heard a noise "like that of a shell." The lights went out, and in the darkness he felt cold water lapping around his legs. Climbing to the hatch, he reached the deck just as the ship sank. A lookout on the deck near the bow was thrown high into the air, but landed in the water, unharmed. The

Lehigh heard an "unusual, but not very loud report," saw a cloud of smoke, and then lost sight of her consort. Men's voices calling for help signaled the beginning of the rescue operation.

At daylight all that remained visible of the 844-ton monitor was the top of her stack. This was noted by the Confederates, who jubilantly announced the sinking. Captain J. A. Simons of the Torpedo Service wired that it had "been my ambition to teach them a lesson," and that the Federal ship had "met a fitting fate." The Secretary of War termed the sinking "gratifying . . . additional evidence of the value of the Torpedo Service," a feeling that was shared by others in varying positions of responsibility.[9]

CHAPTER XVII

SWEEPING THE COAST

When it became apparent that Sherman's army would isolate Charleston, the city and its fortifications were abandoned by the Confederates during the darkness of February 18, 1865. The next morning, Federal soldiers and sailors occupied the smoking ruins. This great citadel, perhaps the best fortified in the Confederacy, a symbol in the minds of Rebels and Yankees alike, did not fall in an attack but because of a simple strategic decision: Sherman feinted an advance, then knifed to Columbia, sealing the Palmetto City from the rest of the country. It fell with all its cannon, forts, and ships.

At wharves, or sunk in the Ashley and Cooper rivers, the Federals found the torpedo boats. Boat *No. 1* was lying at the foot of a railroad wharf, uncompleted. Beside her was a sister ship, and near Chisolm's Mills another lay, "much worm eaten." The water had ruined her boilers and parts of the engines had been removed. Boats *No. 4* and *No. 5* were on the bottom with her in as bad shape, and a larger boat was found near Bennett's sawmill "in good condition." Three other torpedo boats, veterans of extensive service, were soon raised and repaired. One of them was sent to the United States Naval Academy, and another was shipped north on the U.S.S. *Mingoe* but was lost off Cape Hatteras in a storm on June 6, 1865. Some of the remaining ships, including one called the *Midge*, found their way to the Brooklyn Navy Yard.[1]

Soon after he established a base in the city, Dahlgren was approached by several former Confederates who reported that they knew the location of torpedoes in the harbor and, what was more, knew how to raise them, for they had been responsible for putting some of them down. The leader of the group, interestingly enough, was the erstwhile head of the Charleston Torpedo Service, M. Martin Gray. Released from confinement, he evidently offered his services to his former enemies to curry favor. Perhaps Gray also wished to escape the dire punishments with which certain Union commanders threatened anyone who had a part in using torpedoes. The stories he told were both clever and interesting. He claimed, among other things, that he had deliberately sabotaged many of the torpedoes so that they would not explode and damage Federal ships.

Two other members of the defunct service who approached the victors were Francis Wood and Robert Thompson. They indicated the areas where the mines were thought to be anchored, drew diagrams of fuses, and explained how the electric tank torpedoes worked. A model of one of the frame torpedoes was made for Dahlgren by another former Rebel, and the officers studied means of raising and destroying them. Master John L. Gifford was put in charge of these operations and was given several small ships to use in sweeping the harbor.

The Federals began their work on February 19, concentrating on the obstructions off Fort Sumter. Using the tugs *Jonquil* and *Gladiolus,* they took out twenty-two wood obstacles, each fifty feet long, in the first four days. Two floating torpedoes were recovered at the same time, one of them near the wreck of the *Patapsco.* Other channels were dragged on February 26, and one electric torpedo was found. (Gray claimed that when it was put in place two years earlier, he had rendered it harmless.) In March, the scene of the operation was changed to the Cooper and Ashley rivers where frame torpedoes were known to be located. Two more electric tanks with their wires and cables intact were found off Battery Bee, opposite Fort Sumter. At various places around the city, including the old torpedo factory on Hazel Street, the investigators found large numbers of keg and iron torpedoes, and on the wharf was a boiler all ready to be filled and anchored.[2]

Dahlgren was at Georgetown, South Carolina, on the morning of March 1, getting ready to return to Charleston. He was on the *Harvest Moon*, a 645-ton side-wheel gunboat, high-stacked and doubledecked. In his cabin, the admiral paced the floor, awaiting breakfast, and, to pass time, occasionally peered at the shore with his spyglass.

Suddenly without warning came a crashing sound, a heavy shock, the partition between the cabin and wardroom was shattered and driven towards me, while all loose articles in the cabin flew in different directions. Then came the hurried tramp of men's feet, and a voice of someone in the water . . . shrieking. . . . My first notion was that the boilers had burst; then the smell of burnt gunpowder suggested that the magazine had exploded.

I put on a pea coat and cap and sallied forth. Frightened men were struggling to lower the boats. I got by them with difficulty. They heard nothing; saw nothing. Passing from the gangway to the upper deck ladder, the open space was strewed with fragments of partitions. My foot went into some glass. The Fleet Captain was rushing down, and storming about. I ascended the ladder to get out on the upper deck to have a full view of things. A torpedo had been struck by the poor old "Harvest Moon", and she was sinking. The water was coming in rapidly through a great gap in the bottom. The main deck had also been blown through. There was no help for it, so we prepared to leave. . . .[3]

The torpedo had exploded between a starboard gangway and an open passage when the ship was about three miles from Battery White. The *Harvest Moon* was turning along at a good speed because the channel had been cleared of mines. "So much has been said in ridicule of torpedoes," said Dahlgren, "that very little precautions are deemed necessary. . . ." Neglect cost the life of one man and the services of an entire warship—she sank in five minutes.[4]

On March 6 the tug *Jonquil* had been dragging the mouth of the Ashley with a fifty-pound grapnel when a set of frame obstructions with four cast iron torpedoes was snagged. Carefully, the mass was pulled from the river bottom and three torpedoes were secured. The logs supporting the fourth were seen, but the mine could not be located. The tug began towing the other three to shore when the missing torpedo struck the bottom and exploded beneath the midships. The boilers jumped half a foot, and the shock threw nine men overboard and flooded the ship. A man on the berth deck was thrown against the overhead, and his head was split open. Doors were shattered, every window was broken, three beams were seriously sprung, and a howitzer forward was upset. Fortunately the explosion was too far from the hull to penetrate it, and the *Jonquil* was able to go on with her work after repairs.[5]

On St. Patrick's Day the Coast Survey steamer *Bibb*, returning to Charleston after conducting surveys on the bar, received a severe shock at 5:25 P.M. It was described as "like being hit by a rock." A column of water blew up alongside the port bow and fell into the ship's second cutter, tearing it from a davit. Sixty feet of 1½-inch mooring chain coiled on the port side was thrown over the deck, lashing wildly. The staunch little ship rolled heavily, pouring water from her scuppers, but righted and settled herself. Damage parties rushed forward and reported that the injury was slight. Like the *Harvest Moon*, the *Bibb* was hit while cruising waters that had been cleared.

Two days later, the *Massachusetts*, a 1,155-ton navy transport, was mined near the *Patapsco*. She hit a torpedo "heavily" under the starboard quarter, then cut it in half. The mine, probably ripped from its moorings by the ship's keel, did not detonate, but before it could be picked up it sank.

After bypassing Charleston, Sherman marched into North Carolina. On the coast lay Wilmington, the last port on the Atlantic open to the Confederates. Like Charleston, it might have been isolated had not a sea-borne assault caused its capture on January 22, 1865. The action that preceded the fall of the city and, for all intents, destroyed its value to the Confederates was the capture of Fort Fisher which protected the mouth of the Cape Fear River. This great system of sand defenses occupied Bald Head Island and was shaped like the letter L lying on its side. The short or land face extended across the neck of a peninsula for 480 yards and mounted twenty-one guns. Along the sea face for 1,300 yards were sited the most powerful guns, seventeen in all, ranging from

6⅝-inch rifles and 150-pound Armstrongs to 10-inch Columbiads. These guns were all behind twenty-five-foot parapets that were twenty feet thick in most places. The fort fell only after two terrific battles, the last of which included an assault by six thousand soldiers, sailors, and marines. It had been softened by the concentrated shelling of the largest fleet assembled in America prior to 1917. This battle succeeded where the earlier landing of December, 1864, had failed, and has been described often. But one section of the defenses has been little investigated by historians: a line of torpedoes in the sand along the land face. They were outlined in a little-known letter from Brigadier General Cyrus B. Comstock, U.S.A., who had led the land attack, to General Richard B. Delafield, Chief of Engineers, on January 23, 1865:

In front of the land face . . . at an average distance of 200′ from the work, and 80′ from each other, was an elaborate system of torpedoes, 24 in number . . . connected with the fort by three sets of . . . wires. . . . A single wire, running to a group of torpedoes, was branched to each, in the expectation apparently of having battery power sufficient to fire the whole group; and in addition, some of the groups were connected with each other, thus giving . . . a choice of positions in the work to fire the group from.

Shells had cut the sets of wires . . . no torpedoes whatever were exploded. . . . They could not fire those whose wires were uncut, as the fuse I have examined had its powder caked; but it may have been intended for a slow match.

The batteries for firing were magnetic, [with] a few turns of the crank, . . . readily firing gunpowder in fine grains.

The accidental cutting of four of the six wires leading from the work was a piece of good fortune which probably saved us from severe loss and demoralization.

Besides this system, which used the Singer torpedoes fitted out as land mines, a "torpedo house" was found near the famous Mound Battery midway on the sea face, from which wires ran to a galvanic apparatus in the water.[6]

After capturing Fort Fisher the Federals occupied Fort Caswall across the estuary and began to move up the Cape Fear River, carefully dragging the channel. Digging trenches in the sand in front of Caswall, a crew found wires leading to the water. They traced them with grapnels to four large electric mines. Three were raised but the last was left in place after the wires had been cut.

At Fort Anderson, a breastwork constructed on the ruins of Brunswick, an eighteenth-century town about halfway up the river, the sailors noticed wires running into the water from the fort. Tracing the wires, they came upon an example of the most advanced electrical torpedo system the war produced. At the land end was a magneto battery, known as "Wheatstone's Magnetic Exploder," which resembled the transformers on modern toy trains and was of English manufacture. Devised by Matthew F. Maury in England, it consisted of three magnetos in an "elegant" mahogany box. On the box was a crank that, it was claimed, fired as many as twenty-five circuits "with such rapidity that the effect upon the ear is as one explosion." [7]

On the evening of February 20, 1865, Admiral Porter received word that the Confederates were going to make a mass attack against his fleet by floating a hundred mines down the Cape Fear. As a precaution, double lines of fish nets were strung across the river and picket boats were sent out to patrol. Around 8 P.M. a barrel bobbing in the water was spotted by Porter. A boat from the *Shawmut* rowed over and examined it. The boat's commander, Ensign W. B. Trufant, drew his pistol and fired into it. This was an approved method of destroying torpedoes, but Trufant was too close; it blew up, wounding the ensign, killing two of his men, injuring the others, and destroying the boat.

The next night the nets and boats were out again. Some torpedoes were blown up by musket fire (at safe ranges), but one slipped through and caught in one of the *Osceola's* paddle wheels. When it exploded it "knocked the wheelhouse to pieces, knocked down some . . . bulkheads and disturbed things generally." Not long afterward, on March 4, the army transport *Thorne* was sunk just below Fort Anderson. [8]

Later, a contraband appeared beside Porter's flagship, the *Malvern*, and was taken aboard. He told the admiral that a Confederate torpedo boat, the *Squib*, and the gunboat *Yadkin* were nearby, and planned to descend on the fleet under cover of night. Porter immediately ordered every ship to keep a close watch and to maintain a pair of boats in the water ready to board. Each ship was also equipped with a net to foul the torpedo boats' propellers. The lead boat strayed too far, however, and missed the intruders. Porter was readying for bed when he heard shouts in the water, followed by pistol shots and hurrahs. As he ran on deck, lights revealed a swarm of boats. Some five minutes later the *Malvern*'s lookout saw a dark shape followed by the furiously paddling cutters.

"Here he comes!" he sang. A ship incautiously burned a Coston signal flare, lighting up the water.

"We've got him," Porter heard. "Tie on to him!"

"Why didn't those fellows . . . jam his screw with the nets?" the admiral demanded. The *Malvern*'s captain smothered a smile.

"He hadn't any screw, sir."

"Then what had he?" The officer could not restrain his laughter any longer. "It was something worse than a ram; it was the biggest bull I ever saw," reported Porter with humor. "He was swimming across the channel . . . !" [9]

CHAPTER XVIII

OPENING THE JAMES
AND WESTERN WATERS

The Confederate flag-of-truce steamer *Schultz* had taken a group of Union prisoners to Cox's Wharf on the James to be exchanged for captured Confederates. The *Schultz* was a river steamer that had plied these waters before the war, and she made regular trips down to the fortifications at Drewry's Bluff at 3 P.M. each day. Returning to Richmond on February 17, 1865, she was blown up and sunk in the channel off Bishop's Bluff. There was no question that the explosion was that of a torpedo, but whose torpedo was it? On a previous trip the *Schultz* had reported sighting Union mines near Cox's Wharf—how her skipper was able to determine their ownership is not known—but after the ship sank, the official report adjudged that the cause was one of Beverly Kennon's torpedoes which had broken away from its moorings.[1]

Misfortune had beset another river steamer, the *William Allison*, that February; not long before, she had run down the Rebel torpedo boat *Hornet* and sunk her. None of the smaller ship's crew was injured, but the James River Squadron was deprived of the service of a valuable torpedo boat at a critical time. Steps to raise the *Hornet* were begun immediately. The river was dragged and the wreck buoyed, but ice forced a postponement of actual salvage operations. Also, the area was in range of Union cannon, which would certainly do their best to prevent the raising. The sinking left but a single torpedo boat with the squadron—the *Wasp*. The *Torpedo*, which had been used as a steam-tug and mail boat in addition to her services with the Submarine Battery Service, was assigned to the torpedo squadron to make up for the loss.[2]

If the last major move to float a new class of torpedo boat had succeeded, the loss of the *Hornet* would not have been as serious. Plans for a fleet of the boats had begun in July, 1864, when Secretary Mallory finally decided to follow the advice Beauregard had pressed on him two years earlier, and have a number of these new boats built in Europe. He sent Commander James D. Bulloch orders to have six built, because "experience has shown us that, under certain conditions, we can operate effectually against the enemy's blockading fleets with torpedo boats. . . ." Although he enclosed plans and specifications based on the recommendations of several persons, Mallory allowed Bulloch to build three boats as he saw fit. The most important considerations were for the ships to be low and speedy ("at least 10 miles an hour") and to have a crew of between five and seven men. Torpedoes were carried on an A-shaped frame strapped to the bow.

Bulloch studied the plans and made a few changes to strengthen and add more rake to the stern. He figured the boats would weigh 9.18 tons fully equipped. The contracts for three were let in England in September, and full plans for the others were being completed. The latter plans, drawn by J. S. W. Dudgean, were for 60-foot ships with a 12-foot breadth and 7-foot depth, to be powered by two engines. However, the design appeared to be impractical and was canceled. In late January, 1865, six more boats were being designed, but none ever reached America; they were still in the builders' hands when the Confederacy was dissolved.[3]

In the meantime, the torpedo laboratories at Richmond worked to perfect better weapons and to make enough of them to use in nearby battle areas. They were being worked on in several offices. The Navy's Office of Ordnance and Hydrography was turning out many spar torpedoes for its warships, and the Engineer Bureau under Colonel J. F. Gilmer was at work building keg, Singer, and frame models. This was in addition to the work being done by Rains's torpedo bureau and other commands.

At the same time quite a bit of experimental activity was under way. The destruction of most of the records has prevented details from reaching later generations. Even the Federals could find little actual design material when they took Richmond. The only concrete evidence is the few physical specimens that they found.

The devices were a varied and interesting lot. The "devil circumventor" or "turtle torpedo" was a boiler-iron hemisphere that was designed to lie on the bottom, presenting nothing but a curved, smooth surface to grapnels. It was to be sunk beneath a floating torpedo and would explode when the mine above it was removed. The "current torpedo" employed a propeller on the Singer plunger. When revolved by the tide or current, the screw pulled a safety pin and released the spring-driven rod. (The inventors assumed that the propeller would not operate until the torpedo caught onto a ship.) Perhaps the most interesting device to those of the Atomic Era is the "hydrogen torpedo."

This apparatus looked for all the world like a tin washtub with a cover. It depended for explosion "upon the well-known phenomenon of rendering spongy platinum incandescent by throwing on a jet of hydrogen gas," wrote a contemporary expert. A small tank of compressed hydrogen was arranged so that it would flow across a piece of platinum inside a mass of fulminate. The heat would detonate the mercury and then the powder. Whether this weapon was ever used is not known. However, a working model was captured at Richmond.[4]

Hundreds of mines had been planted on the approaches to Richmond, on land as well as in the river. Soldiers marching into the burning city on April 3, 1865, passed through extensive minefields. Rains had personally supervised their placement, reporting in mid-November, 1864, that 1,298 machines had been set out. Additional hundreds were manufactured during the following months. Three feet behind each mine were small red flags on yard-high staffs, which were replaced by dark lanterns covered with red flannel at night. Long streamers of white tape on the ground marked safe paths for Confederate soldiers. These markers were supposed to be removed when the enemy advanced. Union troops were aware of the danger: "On all the roads approaching the city, torpedoes are being laid and covered with dust. . . . Cords 400 feet long are attached . . . and men secreted in the bushes pull the cord on the approach of the enemy."[5] The "cords," of course, were tapes marking the safety lanes that had not been taken up.

The approach to the city by water was just as fraught with danger, and the ships had no recourse but to use the river. Admiral Porter alerted his ships as soon as he learned that Richmond had been evacuated. Several were prepared to go in ahead of the main body to remove the torpedoes. First to attempt the trip were the tugs *Alpha, Watch,* and *Saffron,* and the gunboats *Unadilla, Chippewa,* and *Commodore Perry.* They went upriver in a diagonal formation across the channel, dragging the bottom. On the banks, shore parties moved with them, keeping pace while looking for electrical stations.

We reached the first torpedo . . . a little red flag marks it, by which the boat slips tremulously, . . . Here is a monitor with a drag behind it, which had just fished up one; and the sequel is told by a bloody and motionless figure upon the deck. These torpedoes are the true dragon teeth of Cadmus, which spring up armed men.
Happily the Rebels have sown but few . . . and the position . . . was pointed out by one of their captains who deserted to our side. . . . Great hulks of vessels and chained spars, and tree tops . . . reach quite across the river . . . All along were the deep funnel-shaped cases of the torpedoes just disentombed.[6]

The Rebel captain was R. O. Crowley of the Submarine Battery Service. With him was Edward Moore who as boatswain of the *Patrick Henry* had accompanied Robert Minor on his torpedo attack in Hampton Roads on October 9, 1862. Both men had been active in torpedo units until the evacuation of Richmond. They accompanied the Federal ships and indicated where the Confederates had established the torpedo plantations, as well as pointing out the electric stations. When the explosives were located, red flags were set up on buoys. The shore parties dragged the mines to land and destroyed them.[7]

The torpedoes would prove to be both an aid and a menace. As an aid, they came in handy when the Navy tried to clear the James of the wreckage left by the Confederates. One of these wrecks, the hull of the *Virginia*, lay a mere six feet under water, blocking a good part of the channel. Lieutenant Commander Homer C. Blake decided to destroy her with captured torpedoes. He used several kinds, ranging from one hundred to four hundred pounds, but they had little effect. Next, a more elaborate plan was tried. A trench some thirty feet deep was dredged alongside the *Virginia;* torpedoes were planted at 15-foot intervals on her other side. The simultaneous explosion of all the mines pushed the hulk into her grave.

The weapons were always a menace—often long after their implanting—and bobbed up at unexpected moments. For example, a U.S. torpedo boat hit one in Trent's Reach in December, 1865. The torpedo had been immersed for at least a year, a fact that probably accounts for the detonation of one fuse without touching off the main powder charge.[8]

The work of clearing the James had hardly begun when President Lincoln went up the mine-infested river. He had originally wanted to go to the former Confederate capital by horse, but Porter dissuaded him, bragging that there was "not a particle of danger from torpedoes," since his ships had cleared the channels. When Lincoln saw the gray and black cases on the bank, he remarked to the admiral with a wry grin: "You must have been awful afraid of getting that sergeant's old horse again to risk all this!"[9]

But even in the tragedy of defeat there is sometimes humor. A glimmer lightened the somber train ride that took President Davis and his government away from the Confederate capital. The story is best told in its original form by Burton N. Harrison:

Among the people who beseiged me for permits to get on the train was General Raines, with several daughters and one or more of his staff officers. He had been on duty with the 'torpedo bureau' and had with him what he considered a valuable collection of fuses and other explosives. I distrusted such luggage . . . the General confidently asserted the things were quite harmless. I told him he couldn't go with us—there was no room. . . . He succeeded . . . in gaining access to the President, who had served with him . . . in the [U.S.] Army, and, in kindness to an old friend, Mr. Davis . . . actually [took] the daughter to share his own seat.

In the midst of it all [the delay before the train started] a sharp explosion occurred very near the President, and a young man was seen to bounce into the air, clapping both hands to the seat of his trousers. We all sprang to our feet in alarm; but . . . found that it was only an officer of General Raines' staff, who had sat down rather abruptly upon the flat top of a stove [there was no fire in it] and that the explosion was . . . one of the torpedo appliances he had in his coat-tail pocket.[10]

After the war's end Beverly Kennon wrote Federal authorities giving directions as to where some mines he had buried near Potomac Creek could be found. "Unless [they] are removed," he said, "some innocent body may suffer." He had buried the cases, each with eighty pounds of powder, in the stable yard of a farm and the primers (with "safety tubes on the caps") beside a road nearby. Lieutenant J. H. Eldridge, U.S.N., and a party from the *Delaware* recovered them, but not without difficulty. They turned out to be copper tank buoyant mines with five primers each—a type used during the latter stages of the war.[11]

The fall of Richmond left Mobile as the only port east of the Mississippi not in Federal hands. To be more exact, the harbor was Union, but the city was under Confederate control. After forcing his way into the harbor, Farragut planned to attack the city itself in a joint army-navy operation. It took months to prepare. Ships had to be brought in, supplies laid by, and the army of General E. R. S. Canby had to fight its way to the outskirts. By late March, 1865, all was ready.

In Mobile itself, the Confederates had been actively preparing for the battle. General Dabney H. Maury, a nephew of the "Pathfinder of the Seas," was in charge, and one would have to look far for a more energetic leader. "Old Puss in Boots" his men called him, for he was not very tall, and half of him seemed to be swallowed in the pair of thigh-length cavalry boots he favored. With a bristling black beard covering his features, Maury made up for his lack of stature by action. Every approach to Mobile was protected by ditches, obstacles, heavy guns, and torpedoes. Every channel and anchorage was sown with mines and each road and battery was surrounded. This had been accomplished under the direction of Maury's chief engineer, a German mercenary, Lieutenant Colonel Victor von Sheliha.

A "scientific officer," Sheliha was a graduate of Prussian military schools, had served as a Lieutenant in the Sixth Infantry Regiment of the Prussian Army, and was living in New Orleans in 1861. He joined the Confederate Army as an engineer in General F. K. Zollicoffer's command that same year, was captured at Island No. 10 and imprisoned at Fort Warren, Massachusetts, in May, 1862. When exchanged on August 13, 1863, he reported to General Simon B. Buckner. On October 10, he was ordered to Mobile. Sheliha was extremely articulate and wrote good English in a fine, legible script. He was one of the most interesting persons engaged in torpedo warfare and, happily for posterity, set down many of his experiences in a little-known book, *A Treatise on Coast Defence*, which was published in England in 1868.[12]

Maury still had one torpedo boat. She was the *St. Patrick*, a David which had been built for the government by John P. Halligan who was to skipper her as well. She was due to be ready in July of 1864, but her completion was delayed and Maury anxiously pressed Halligan, hoping to be able to use her when Farragut attacked. When Halligan finally did tow the vessel to Mobile, he seemed to stall and make up excuses as to why he should not attack. Maury's patience grew thin and, finally, he threatened to replace Halligan with an officer who would take the David into battle. He told Commander Ebenezer Farrand, C.S.N., the same thing and, when nothing was done, arranged to have the ship transferred to the Army. Lieutenant John T. Walker, C.S.N., replaced Halligan, but when Walker boarded the *St. Patrick* he found that Halligan had absconded with certain vital parts of her machinery. Tracing the missing contractor, they found him "comfortably established" at the Battle House Hotel in Mobile. The machinery was recovered "by energetic and good management," as Maury put it.[13]

The *St. Patrick* was "shaped like a trout," fifty feet long, six feet wide, and ten deep. Her copper torpedo was on a twelve-foot pole and the vessel could be lowered in the water, as could other Davids, by filling her ballast tanks.[14] She was ready on January 27, 1865, and that night she steamed down the bay and into the enemy fleet. Between 1 and 2 A.M. on the twenty-eighth, she began her run, selecting the gunboat *Octorara* for her target. The Federal sighted her while she was still some distance astern.

"Boat ahoy!"

"Aye, Aye!"

"Lie on the oars," sang out the Yankee, mistaking the David for a cutter. By this time the *St. Patrick*, a bit off course, rasped alongside the gunboat. As she passed, the captain of the *Octorara*'s deck guard reached out in desperation, wrapped an arm around the Rebel's stack, and shouted for his men to tie her with a rope. Below, on the torpedo boat, the Confederates began shooting at the Yankee overhead and forced him to release the hot pipe. The *St. Patrick* disappeared in a hail of gunfire and returned to Mobile practically undamaged. The war was over before she made any more attacks.[15]

Sheliha had experienced great difficulty in getting the static defenses ready. A sharp, irritable person, plagued by a "chronic disease of the liver," the Prussian had little patience with delay. During August, 1864, he complained that "precious time is being lost for want of labor" and appealed to the governor of Alabama and to General Maury "to procure hands." He threatened to resign immediately unless the government backed him. When help was not forthcoming as quickly as he felt it should be, true to his word, Sheliha sent in his resignation. It was refused.

As the year came to a close, his health became increasingly worse, and on December 31, he requested six months' leave to go to Europe to recuperate at "the Springs in Bohemia." His request was granted this time, and the work of supervising the torpedo defenses at Mobile fell to Lieutenant J. T. E. Andrews.[16]

On March 12 one of the U.S. warships, the tiny 72-ton, screw-propelled gunboat *Althea*, mounting a single cannon, was dragging the channel with a chain attached to a spar laid across her stern. Off Battery Ruger, the chain fouled on an old wreck. Efforts to free it failed and it was "slipped." Right afterward the ship was shattered by an underwater blast on her port side, behind the pilot house. The *Althea* went to the shallow bottom like a stone. Two of her sailors died and three were hurt, including Ensign F. A. G. Bacon, her captain.[17]

The main obstacle to a Union attack on Mobile was an earthwork known as "Spanish Fort." It lay on the east side of Mobile Bay in a tongue of land that commanded a tortuous channel. Actually, the fort was a battery of six heavy guns and a garrison of some 2,100 men. A total of 205 Rains subterra shells blocked the land face, and in the harbor were several hundred Singer torpedoes.[18]

Admiral Henry K. Thatcher, U.S.N., made his dispositions on March 28 and paraded four monitors—the *Winnebago,* the *Osage,* the *Kickapoo,* and the *Milwaukee*—and the gunboat *Octorara* while a military column moved out from Pensacola, Florida, via Blakely to the fort. That night the squadron commenced its bombardment of Spanish Fort as the army made its demonstrations on the land side.

The next afternoon the *Winnebago* and the *Milwaukee* were sent up the Blakely River, which ran beside the fort, to shell a Confederate transport carrying supplies to the garrison. When that vessel moved farther upstream, they received orders to return to the fleet. They drifted back with the current, using just enough steam to "keep the head up" against the tide. "My object," wrote Lieutenant Commander James H. Gillis, of the *Milwaukee,* "being to avoid in turning, the accident that caused the sinking. . . ." A torpedo exploded under her port side, forty feet from the stern. The fantail of the 970-ton, double-turreted "river monitor" settled in minutes, but the bow did not fill for nearly an hour. Her skipper said: "There was naturally some confusion at first, the hatches being closed, [for action] and but three [men] being provided with levers to open them from below, and those who were not on deck being dependent on those who were, for other means of egress; but a single command served to restore order, and all came on deck in a quiet, orderly manner." [19]

The *Milwaukee*'s sinking was especially surprising to the Gulf Blockading Squadron since this stretch of water had been cleared of torpedoes the night before. But Gillis was not completely despondent. After divers had been sent from Pensacola, he said that he felt "there is every prospect of my retaining my old command until I have the pleasure of seeing her guns once more used against those who were no doubt now exulting over her supposed loss." [20]

The next afternoon (March 29) the *Winnebago* was blown away from her anchorage by a strong east wind and came close to colliding with the *Osage,* a monitor with two guns in a forward turret and an iron turtleback covering her paddlewheels. To prevent a collision the *Osage* raised her anchor and moved ahead a bit. Her skipper, Lieutenant Commander William M. Gamble, got ready to anchor in the new location. At 2 P.M. he rang three bells to back her, and was standing at the forward door of the pilot house preparing to mount the turret and supervise the operation. He had taken less than three steps when an explosion beneath the bow shook the vessel and poured water over her forward deck. The 523-ton monitor began to go down in twelve feet of water. Five of her crew were killed and eight wounded by the blast.

If the loss of the first monitor had been a shock to the fleet, the sinking of the *Osage* was more so, for the place where she went down had been very carefully searched after the loss of the *Milwaukee*. The cannon of Spanish Fort and of a blockade runner with the favorite name *Nashville* saluted the success.[21]

Salvage efforts were begun at once. At 1 P.M. on April 1, the *Rodolph,* a small, stern-wheeled, tinclad gunboat with six cannon, picked up a barge load of salvage gear for the *Milwaukee* and towed it toward the hulk. She never completed her mission, however, for at 2:40 that afternoon, midway between the *Chickasaw* and the *Winnebago,* a torpedo blew a ten-foot hole in her bottom. The *Rodolph* sank rapidly, taking four of her complement with her. Nearly a dozen men were wounded.[22]

Admiral Thatcher himself had a narrow escape on March 29. A drifting mine, scooped from the bay by one of the gunboats, had supposedly been emptied of its contents and was sent to the flagship for his inspection. At his bidding a couple of sailors were unscrewing the percussion nipples when one exploded. The blast disintegrated the torpedo, wounded two seamen, but miraculously missed the admiral who sat in a deck chair less than five yards away.[23]

Ashore, the siege of Spanish Fort continued until 1 A.M. on April 9, when the fort was surrendered. As the Federals advanced, they came upon the land mines. These were a problem, one Union officer thought, more because of their effect on the men's morale than their actual power of destruction.

These . . . appeared to be 12-pounder shells, filled with powder, furnished with a metallic plug similar in shape to the common fuse-plug, through which passed a needle, the latter falling upon a nipple with the ordinary percussion cap. They were placed upon all approaches to the rebel works, and in every path over which our troops would be likely to pass. Even the approaches to the pools of water, upon which the men relied for cooking, were infected with them. On the roads, they were generally placed in the ruts. As their explosion depends entirely upon their being stepped on, very few of them were effective, and the cases, in which men, horses, or wagons were injured were isolated. Still, the knowledge that these shells were scattered in every direction would necessarily produce its effect upon the troops, who never knew when to expect an explosion, or where to go to avoid one.

The carelessness evinced by the Rebels, in marking the places of their deposit, is most culpable, as many of them could not be found, are liable at any time to injure persons, who from curiosity, or other motive, may visit the ground.[24]

The soldiers themselves appear to have stoically accepted the risk of exploding the mines. "The Rebs," wrote H. U. Dowd of Company A, 114th Ohio, after the attack, "had torpedoes planted in the Ground. Killed several men."

To retrieve the explosives, the Union commander did as Sherman and McClellan had done and put his prisoners to work. "All quiet today," noted Dowd in his pocket diary on April 10. "Troops laying in Camp. Got the Rebs digging up their torpedoes around the post at Blakely. . . ." One of the mines was overlooked, and on April 18: "Nothing going on today of note . . . the Boys getting anxious to go north. . . . One man got his leg Blowed off by a Torpedo." [25]

A total of 150 torpedoes had been recovered from the bay and its tributaries by April 12, a service which Admiral Thatcher felt was "demanding in coolness, judgement and perserverance." Commander Pierce Crosby directed this harrowing work from the gunboat *Metacomet*. To "sweep" the Blakely River, Crosby arranged to use large nets suspended across the channel between twenty rowboats. Moving up and down the waterway, they swept one area six times, then crossed from shore to shore twice on April 9. The booty was twenty-one large floating mines.[26]

On the morning of April 13, the tug *Ida*, going to report to the *Genesee*, struck a torpedo that burst her boilers and completely destroyed her. A seaman on the *Albatross* saw her explode: "I think, her smoke stack must have gone fifty feet into the air. There seemed to be a thick mist about her, hiding her completely from sight. When it . . . cleared . . . she had sunk, but as the water was shoal . . . her upper deck was out of the water." [27]

"We dreaded torpedoes more than anything else," this same sailor reported. Admiral Thatcher phrased it differently: "These are the only enemies that we regard." They both were right in their apprehension, for the very next day the wooden gunboat *Sciota* ran onto another of the deadly contrivances. She had finished coaling from a barge sent by the brig *American Union*, and was delivering working parties to various ships when, on her way to the *Elk*, she was lost. "The explosion was terrible," said her commander, "breaking the beams of the spar deck, tearing open the waterways, ripping off starboard forechannels and breaking the foretopmast." Five were killed, six wounded, and the ship was a total loss. [28]

On the same day a launch from the *Cincinnati* was dragging the channel. One of the explosives was hooked and "weighed" to within two feet of the surface when the mooring broke, causing the drag rope to snap the torpedo against the cutter's stern. The explosion destroyed the craft and killed three of its crew. [29]

This was the last torpedoing in Mobile Bay during the war. Mobile was captured on April 13 and fighting subsided in the South, but the mines were no respecters of agreements, and another ship fell victim before the harbor was cleared. She was the *R. B. Hamilton*, a 400-ton army transport which was destroyed on May 12, 1865, with part of the Third Michigan Cavalry aboard. Thirteen men were killed or wounded. [30]

The toll of mines at Mobile was great—far greater than anyone had ever anticipated. Though the sinkings came too late to have any effect on the outcome of the siege, they indicated the great damage a bay full of torpedoes could create. Since Farragut's attack of August 5, 1864, nine warships and a launch had been sunk, killing or wounding some 200 seamen. These losses included all kinds of ships, from the most powerful monitors to wood tugs and transports.

Even after the surrender of Mobile, there was a portion of the Confederacy that remained unconquered, with armies in the field and a navy afloat. Including Texas, Arkansas, West Louisiana, and the Indian Territory, this was the Trans-Mississippi, commanded by General E. Kirby Smith. The ironclad *Missouri* lay up the Red

River, and at Galveston, a port still open to the Gulf, were several small ships plus what the Federals heard was a torpedo boat "shaped like a box." Actual war with mines here was sporadic after mid-1864. The U.S. Navy maintained firm control over the rivers in its possession and kept Kirby Smith's command separated from the rest of the Confederate states. The only communication was by couriers, many of whom were caught.

Ships in the Mississippi began reporting floating torpedoes once more in August, 1864. They were recovered and found to be wicker-covered demijohns like the ones that sank the *Cairo* in 1862. These devices were made at Black Hawk Point and near Bayou Sara, Louisiana, several miles from the river, by a group of Confederates under the command of a Colonel Hill. Operating in enemy territory as they were, the Rebels were forced to move in strict secrecy. Their first attack was against the *Lafayette*. It failed when the percussion arrangement of the mines refused to work. They tried again, with the monitor *Ozark* as a target, on September 10. The torpedoes were hauled to the river by the slaves of a local planter and set adrift. They too failed, and the crew began designing other methods of ignition.

Little is known of this operation except what information a Union spy was able to gather by mingling with the Rebels. None of the men were caught, with the exception of a soldier who had been assigned to a unit protecting the factories. The steady pressure brought to bear led to their withdrawal in December. Evidently the group was attached to the Mobile torpedo corps.[31]

The hunt was on again in March, 1865. It was prompted by a telegram on March 8 from Grant to General J. M. Palmer, head of the U.S. Army in Kentucky: "Information from Richmond indicates that a naval party have gone to the Ohio River for some mischievous purpose. Look out for them, and . . . hang them up as fast as caught."

Supplementary information indicated that the party was supposedly ordered by the C.S. Navy to disrupt shipping on the Mississippi. Then a Union steamboat pilot from Chapman's Landing near Kingston on the Tennessee became suspicious of the way the women of the neighborhood "were moving around." He thought they were evidence "of some rebel movement . . . on foot," and decided to find out what was going on. Taking his shotgun, he went down to the river and made "a startling discovery." There, hidden in the bushes, was a large yawl full of boxes. Returning with a hastily organized posse, he captured and disarmed nine men "by issuing orders to imaginary troops."

In the 36-foot boat were boxes of torpedoes, fireballs made of cotton soaked in turpentine, and, "the most dangerous article of all," hand grenades wrapped in cotton. The prisoners belonged to the Confederate Navy, and on January 3 had been sent from Richmond to Bristol, Virginia, with the boat by rail. The yawl had been floated in the Holston River and was headed for Chattanooga with orders to burn ships, depots, and wharves.[32]

The final torpedo operation the Confederates were able to muster came from that last tiny naval squadron hiding far up the narrow Red River. The potential danger of an attack from the Rebel ironclads remained, but was it possible? Officers of the U.S. Navy thought so and ordered vigilance "to run down or board torpedo or other craft."

The Navy went up the river in June to accept the surrender of the Confederate ships, the last demonstration of the war. They proceeded cautiously, having been instructed to "take prompt and efficient means for their [torpedoes] removal." None were found, and the surrender of the C.S.S. *Missouri* with her consorts the former U.S.S. *Champion* and the *Cotton* was accepted.[33]

CHAPTER XIX

AFTERMATH

The surrender of Confederate armies and ships ended all further development of mines and torpedoes by those forces. However, at the time of the surrender Matthew Fontaine Maury was on the high seas bound for Galveston with his new electric torpedo system. In May his ship put into St. Thomas, Virgin Islands, and he learned of the dissolution of the Confederacy. Instead of surrendering, he offered his services to Maximilian in Mexico, and served in the "Office of the Imperial Administration of the Admiralty" for a short time. While there, he released his rights to the torpedo patents to one Nathaniel John Holmes of London, England, who took out patents in his own name on December 8, 1865.

At least two complete sets of Maury's mines had been used in America during the war. These were the elaborate devices at the mouth of the Cape Fear River and off Fort Anderson. The disposition of the set Maury had with him aboard ship has not been discovered. Presumably, it was taken to Mexico. This was without question the most advanced system of mines in the world at that time. Its intricacies and certainty of explosion were minutely described in the patent. It required the decision of two operators to explode any of the mines. The men were placed at right angles so that by sighting along their ranges they could see with certainty whether or not a ship was in the proper position. Maury provided a device to check the circuits and, at the same time, allow the operators to telegraph one another through the mines without risking an explosion. All of this was accomplished by a low voltage "tension" current that passed through tiny platinum wires in the mines. For detonation, much stronger "accumulated" charges were used to ignite the fuses.

By using the Wheatstone detonator, the system could employ a single or a dozen torpedoes, and they could be exploded in any combination on land or in the water. If they were in a channel used by friendly ships, the weapons were anchored by hollow iron shells. To bring them to combat depth, the operators simply pressed a key that ignited a tiny charge in the shell and broke the mooring. Maury designed his cases in an ellipse, "the action of the water upon this inclined plane being similar to that of the wind in raising a kite." Of his new system he wrote: "It is, by the new combinations . . . that I have reduced the torpedo to a system, and established that the various material . . . to build up the mine can all be constructed upon an ascertained and well-defined principle. The novelty of my improvement consist in the combinations . . . to bring about the testing and speaking through the mines . . . and . . . to ensure the ignition . . . only when the enemy shall be within its area of destruction. . . ."[1]

Maximilian's government was another lost cause, and in 1866 it collapsed. Maury eventually returned to the United States and spent the rest of his life on the faculty of the Virginia Military Institute. His remaining years were filled with awards, but those most remembered by his family were the LL.D. from the University of Cambridge and a bronze chest with three thousand guineas from the British government.

Hunter Davidson was also in England when the war ended. He too returned and cast about for a new career. In the spring of 1867 he was at Annapolis, Maryland, near the school he had attended twenty years earlier. He had been unable to find suitable employment and was contemplating joining the Prussian Navy as a torpedo specialist. Later, he served at least two governments in South America: those of Venezuela and Argentina. All this time he engaged in a running argument with Jefferson Davis as to the history of the torpedo service. Davis mentioned neither Davidson nor the services of the group in his book, but he included the exploits of Rains and Maury. Fortunately, Davidson took exception and chronicled his duties and contributions in several letters and articles which shed valuable light on the story of the Confederate torpedo organizations.[2]

Gabriel Rains retired for a time to Augusta and Atlanta, Georgia. From 1877 to 1880 he was employed as a clerk in the United States Quartermaster Depot at Charleston, a job which was far below the position he held in the military for so many years. One son, Sevier McClellan Rains, born in 1851, was graduated from West Point in the class of 1876 and was tragically killed a year later at Craig's Mountain, Idaho, in action against hostile Indians. His son's early death, as well as Rains's own continued disability from wounds he had received during the war, contributed to his death on September 6, 1881, at Aiken, South Carolina. He was seventy-eight years old.

James H. Tomb never returned to the sea. Having served on the Davids, aboard the ironclads, and with a land mine crew, he went to St. Louis after the war and was engaged in the hotel business for forty years. In 1874 he was the proprietor of the Mona House on North Sixth Street and, as late as 1906, ran the Benton Hotel, on Pine Street.[3]

William T. Glassel was made a prisoner of war after attacking the *New Ironsides* with the *David* in 1863. He was exchanged and, in April, 1865, was the commander of the Confederate ironclad ram *Fredericksburg* in the James River Squadron. It was his painful duty to see that the ship was scuttled when Richmond was evacuated. Later, he was offered a position of high rank in a foreign navy on the recommendation of former Commodore John H. Tucker, but he declined. Instead, he spent his life in Los Angeles in quieter pursuits. Glassel died there on January 28, 1879.[4]

Isaac N. Brown, responsible for the destruction of the first warship ever sunk by a torpedo during combat, was active in several theaters: he was skipper of the ironclad *Charleston,* and at the end of the war he was placed in command of all naval defenses west of the Mississippi. The surrender of the armies found him in Texas on his way to the new post. He retired to his plantation in Mississippi "without a dollar." A valiant struggle against disfranchisement and poverty occupied the remainder of his life. He triumphed over both and spent his last years in Corsicana, Texas.[5]

As soon as the war ended, Francis D. Lee, inventor of the torpedo ram, who fathered two types of modern warships—torpedo boats and torpedo-boat destroyers (generally known by only the last word today)—went to France on the invitation of Emperor Napoleon III. Napoleon had heard of his torpedo and boats and was interested in having a fleet for his navy. The two men had half a dozen interviews but could not reach a satisfactory agreement. Next, Lee had conversations with the Admiralty in London, also without positive results. He returned to the United States and settled in St. Louis. There he resumed his work as an architect "and soon rose to the foremost rank of his profession." Several buildings that he designed are still in use.[6]

Undoubtedly James Tomb was a frequent visitor, and one can imagine the conversations they had. "His house became the headquarters for all South Carolinians who passed through that city, and those who have once been entertained by him can never forget his grand hospitality," said a friend. In August, 1885, Lee died of apoplexy at fifty-eight. He was buried in St. Louis.[7]

If any of the people who used torpedoes were liable to the hanging which several prominent Union commanders had threatened, none would have been a better candidate than former Captain Thomas E. Courtenay, founder of the Confederate Secret Service Corps. In 1864 he prudently joined his family in Halifax, Nova Scotia, after Admiral Porter captured his correspondence to his western agents. This was done as much to protect his family as himself, for they had been living in Maryland. From Halifax they went to London, and for five years Courtenay tried to sell the coal torpedo to the British government. He returned to the United States in the early 1870's and died on September 3, 1876, near Winchester, Virginia. He was only fifty-four.[8]

Beverly Kennon, who had done so much to pioneer the development of mines, eventually found a use for his talents. He went to sea after the war and served as an officer aboard an American merchantman. In 1869, he was appointed a colonel in the coast defenses of the Egyptian Army—a position he owed to a New York Militia officer and a rear admiral in the U.S. Navy.[9]

Not one of the operators or inventors was ever punished for having dealt with torpedoes. In fact, some joined their former enemies and demonstrated the secrets. Throughout the war a group of Federal army officers had been collecting and studying the weapons. After the resumption of peace, they gathered their findings for publication. This project was entrusted to the supervision of Major General Richard Delafield, Chief of Engineers. Captains Peter S. Michie and W. R. King examined each type of mine thoroughly, made detailed diagrams, and compiled the data in the book *Torpedoes: Their Invention and Use, From the First Application to the Art of War to the Present Time,* published in 1866. It was considered "of a confidential character not intended for general circulation but for professional use solely." King hoped the study would "be instrumental in bringing this great auxiliary in modern warfare to a degree of perfection commensurate with its importance."

There would be other studies and other developments. The United States experimented with mines and incorporated both floating torpedoes and torpedo boats into its armed forces. Soon Confederate pioneering was overshadowed by startling new developments: the Whitehead self-powered torpedo which led to those used today; submarines propelled by deisel and electric engines as perfected by J. P. Holland; great fields of floating mines; and acres of land mines.

The weapons invented and devised by the various Confederate torpedo commands did not contribute materially to the grand strategic effects of the war. Their physical contributions were on the local level. Nevertheless, coping with them demanded great daring and perseverance on the part of the Federals.

As in all civilized nations, there were many visionaries, planners, and local inventors who freely volunteered ideas for weapons they believed would destroy a ship, an army, or a fleet. Usually there was some vital link missing: the ability of the device to withstand handling or natural elements; an effective means of delivery; or a reliable manner of escape for the operators. Once these inventions were disposed of, there were relatively few practical weapons left for the military and naval authorities to consider.

Essentially, they were divided into two types: defensive, as characterized by the anchored Singer and Rains models; and offensive, such as the drifting ones and spar torpedoes carried on ships. At first they were the products of isolated endeavors and their use was infrequent and sporadic. But the success of the first mines led, as we have seen, to the establishment of agencies which designed and perfected the use of torpedoes. Though propaganda was not used as a weapon of war at the time, the hidden explosives' value in that area was nearly as great as in their intended purpose. Certainly fear on the part of many officers and men had an effect upon several movements, attacks, and maneuvers of Federal ships and troops. The minefields in Charleston Harbor were important in keeping the Union fleet outside; and it was torpedoes, in combination with skillfully planted shore batteries, that turned back several Union attacks up the James and Roanoke rivers. Perhaps these assaults would have been repulsed by cannon alone (as was the first Federal movement up the James in 1862), but the mere presence of mines gave the Confederates an added advantage and forced the enemy to take pains to clear the waterways ahead of him. At Mobile, Farragut was forced to take the torpedoes into consideration, and the loss of the *Tecumseh* unnerved his battle line at a crucial moment. Only his fierce personal determination averted a Federal disaster.

Nowhere did Confederate mines and torpedoes actually change the outcome of a major battle; but at various places they permitted the Confederates to achieve offensive capabilities in essentially defensive operations, and in some actual encounters of major importance, mines delayed the decision sufficiently to give the Confederates time to regroup or escape. For example, off Charleston and Hampton Roads, the Federals cruised at will during daylight, but after nightfall, they gathered their capital ships behind circling boats, log and rope obstructions, and calcium lights. At Williamsburg and Fort Wagner, land mines assisted in slowing Federal advances so that the Southerners escaped completely.

The great Union ship losses to mines occurred late in the war after the outcome was inevitable. Most of the mines and torpedo rams were effective only in the smooth, shallow waters of upper rivers and inner harbors. Losses did not take place until the Federals ventured into those areas. Union losses were therefore in direct proportion to Union attacks.

Offensive Confederate torpedoes were used sparingly. Mines set adrift in currents or tides had limited chances of success. Only the Davids presented an offensive weapon that appeared to be blessed with greater capabilities, and seagoing ships of this type were being planned in 1865. Had the Confederacy lasted another year, these vessels might have taken a prominent role in sea warfare. The same was true of defensive mines. Maury's great electrical systems were too late, and the crude mechanical torpedoes were effective only when used in shallow, narrow areas.

The efforts of the Confederacy proved that a small, poor, and predominately agricultural area could develop fairly inexpensive weapons of a new order. This impressed other nations, who were quick to adopt the weapons themselves. In fact, electrical minefields based on Maury's design, and monitors with few basic changes were used as recently as World War II; and the mines, submarines, torpedo boats, destroyers, and minesweepers in modern navies are results of the crude weapons developed by the Confederates.

Occasionally a voice was raised reminding the world of Confederate pioneering in the development of these weapons, but few listened. Perhaps the most fitting tributes were written by Union officers:

The torpedo is destined to be the least expensive but most terrible engine of defense yet invented. No vessel can be so constructed as to resist its power; . . . the knowledge that a simple touch will lay your ship . . . helpless, sinking . . . without even the satisfaction of firing one shot in return, calls for more courage than can be expressed, and a short cruise among torpedoes will sober the most intrepid disposition. . . .

Another agreed:

Notwithstanding the imperfections and consequent failures . . . the . . . list of vessels destroyed . . . is a sufficient evidence of their utility; and . . . it must be admitted that time, materials, and labor bestowed upon them was well expended. There is but little doubt that with a more perfect system . . . southern ports would have been safe from any naval attack. . . .[10]

UNION SHIPS SUNK OR DAMAGED
BY CONFEDERATE TORPEDOES

DATE	VESSEL	LOCATION	SIZE IN TONS	DAMAGE
Feb. 14, 1862	*Susquehanna's* launch	Wright River, Ga.	——	Minor
Dec. 12, 1862	*Cairo,* armored river gunboat	Yazoo River, Miss.	512	Sunk
Feb. 28, 1863	*Montauk,* monitor	Ogeechee River, Ga.	844	Serious
Mar. 14, 1863	*Richmond,* screw sloop	Port Hudson, La.	1,929	Minor
May 7, 1863	*Weehawken,* monitor	Charleston Harbor, S. C.	840	Minor
July 13, 1863	*Baron De Kalb,* armored river gunboat	Yazoo River, Miss.	512	Sunk
Aug. 5, 1863	*Commodore Barney,* river gunboat	James River, Va.	513	Serious
Aug. 16, 1863	*Pawnee's* launch	Light House Inlet, S. C.	——	Sunk
Aug. 16, 1863	*Pawnee,* gunboat	Light House Inlet, S. C.	872	Minor

Appendix

DATE	VESSEL	LOCATION	SIZE IN TONS	DAMAGE
Sept., 1863	John Farron, army transport	James River, Va.	250	Serious
Oct. 5, 1863	New Ironsides, ironclad	Charleston, S. C.	3,486	Serious
Feb. 17, 1864	Housatonic, screw sloop of war	Charleston, S. C.	1,240	Sunk
Mar. 6, 1864	Memphis, screw gunboat	North Edisto River, S. C.	791	Minor
Apr. 1, 1864	Maple Leaf, army transport	St. John's River, Fla.	508	Sunk
Apr. 9, 1864	Minnesota, frigate	Hampton Roads, Va.	3,307	Minor
Apr. 15, 1864	Eastport, armored river gunboat	Red River, La.	700	Sunk
Apr. 16, 1864	General Hunter, army transport	St. John's River, Fla.	460	Sunk
May 6, 1864	Commodore Jones, river gunboat	James River, Va.	542	Sunk
May 9, 1864	H. A. Weed, army transport	St. John's River, Fla.	290	Sunk
June 19, 1864	Alice Price, army transport	St. John's River, Fla.	320	Sunk
Aug. 5, 1864	Tecumseh, monitor	Mobile Bay, Ala.	1,034	Sunk
Aug. 9, 1864	Ammunition transport	City Point, Va.	——	Sunk
Aug. 9, 1864	Lewis, supply ship	City Point, Va.	——	Sunk
Nov. 27, 1864	Greyhound, army transport	James River, Va.	900	Sunk
Dec. 7, 1864	Narcissus, tug	Mobile Bay, Ala.	101	Sunk
Dec. 9, 1864	Otsego, gunboat	Roanoke River, N. C.	974	Sunk
Dec. 10, 1864	Bazely, tug	Roanoke River, N. C.	55	Sunk
Jan. 15, 1865	Patapsco, monitor	Off Charleston, S. C.	844	Sunk
Feb. 20, 1865	Shawmut's launch	Cape Fear River, N. C.	——	Sunk
Feb. 21, 1865	Osceola, gunboat	Cape Fear River, N. C.	974	Minor

DATE	VESSEL	LOCATION	SIZE IN TONS	DAMAGE
Mar. 1, 1865	*Harvest Moon,* gunboat	Georgetown, S. C.	546	Sunk
Mar. 4, 1865	*Thorne,* army transport	Cape Fear River, N. C.	403	Sunk
Mar. 6, 1865	*Jonquil,* gunboat	Ashley River, S. C.	90	Serious
Mar. 12, 1865	*Althea,* gunboat	Blakely River, Ala.	72	Sunk
Mar. 17, 1865	*Bibb,* coast survey steamer	Charleston, S. C.	——	Minor
Mar. 19, 1865	*Massachusetts,* navy transport	Charleston, S. C.	1,155	Minor
Mar. 28, 1865	*Milwaukee,* monitor	Blakely River, Ala.	970	Sunk
Mar. 29, 1865	*Osage,* monitor	Blakely River, Ala.	523	Sunk
Apr. 1, 1865	*Rodolph,* armored river gunboat	Blakely River, Ala.	217	Sunk
Apr. 13, 1865	*Ida,* tug	Blakely River, Ala.	104	Sunk
Apr. 14, 1865	*Sciota,* gunboat	Mobile Bay, Ala.	507	Sunk
Apr. 14, 1865	*Cincinnati's* launch	Blakely River, Ala.	——	Sunk
May 12, 1865	*R. B. Hamilton,* army transport	Mobile Bay, Ala.	400	Sunk

Chapter 1

1 *Harper's Weekly,* July 27, 1861; *Official Records of the Union and Confederate Navies in the War of the Rebellion* (Washington, 1894–1927), Ser. I, Vol. IV, 566–67, hereinafter cited as *Navy Records* (unless otherwise indicated, all citations are to Series I).

2 Diana Fontaine Maury Corbin, *A Life of Matthew Fontaine Maury* (London, 1888), 192.

3 *Ibid.* These sentiments were expressed by Maury on October 29, 1861, to Grand Admiral Constantine of Russia, who had invited him to take up residence in that country.

4 Richard L. Maury, "Notes by Colonel Richard L. Maury," *Southern Historical Society Papers,* XXXI (1903), 326–28; Corbin, *Matthew Fontaine Maury,* 189–201; Dunbar Rowland (ed.), *Jefferson Davis, Constitutionalist, His Letters, Papers and Speeches* (10 vols.; Jackson, Miss., 1923), VII, 391 n; interview with Mrs. N. M. Osborne, granddaughter of Maury, August 3, 1959.

5 Diary of Betty Herndon Maury (MS in Maury Papers, Manuscript Division, Library of Congress), July 10, 1861, hereinafter cited as Maury Diary.

6 *Ibid.*

7 John W. Wayland, *The Pathfinder of the Seas: The Life of Matthew Fontaine Maury* (Richmond, 1930), 117.
8 *Navy Records*, VI, 26; J. Thomas Scharf, *History of the Confederate States Navy From Its Organization to the Surrender of Its Last Vessel* (New York, 1887), 750, hereinafter cited as Scharf, *Confederate Navy*.
9 Maury Diary, July 8, 1861; Maury, "Notes by Colonel Richard L. Maury," 329.
10 *Navy Records*, VI, 304a.
11 *Ibid.*, 304a–304b.
12 *Ibid.*, 363, 393; *Harper's Weekly*, November 2, 1861.
13 *Navy Records*, XXII, 791.
14 *Scientific American*, March 8, 1862; Frank Moore (ed.), *The Rebellion Record: A Diary of American Events With Documents, Narratives, Illustrative Incidents, Poetry, etc.* (11 vols.; New York, 1864–68), "Incidents," II, 69.
15 *Navy Records*, XXII, 806.
16 *Ibid.*, 807; Royal B. Bradford, *History of Torpedo Warfare* (Newport, R.I., 1882), 48.
17 *Navy Records*, XXII, 651; *Harper's Weekly*, March 29, 1862; Benson J. Lossing, *Pictorial History of the Civil War* (3 vols.; Philadelphia, 1866), II, 237; Scharf, *Confederate Navy*, 751.
18 *Scientific American*, April 5, 1862; Lossing, *Pictorial History*, II, 237.

Chapter II

1 La Fayette C. Baker, *History of the United States Secret Service* (Philadelphia, 1867), 77.
2 *The War of the Rebellion: A Compilation of the Official Records of the Union and Confederate Armies* (Washington, 1880–1901), Ser. I, Vol. XI, Pt. 3, p. 487, hereinafter cited as *Official Records* (unless otherwise indicated all citations are to Series I); Corbin, *Matthew Fontaine Maury*, 200; Maury, "Notes by Colonel Richard L. Maury," 329.
3 James D. Richardson (ed.), *A Compilation of the Messages and Papers of the Confederacy Including the Diplomatic Correspondence 1861–1865* (2 vols.; Nashville, 1905), I, 204.
4 Hunter Davidson, "The Electrical Submarine Mine 1861–1865," *Confederate Veteran*, XVI (1908), 456–57; *Navy Records*, VII, 545.
5 *Navy Records*, VII, 543–44.
6 *Ibid.*, 780.
7 *Ibid.*, 543–45.
8 *Ibid.*, 546.
9 *Ibid.*
10 *Ibid.*, 545; Rowland (ed.), *Jefferson Davis*, VII, 109.
11 Hunter Davidson, "Electrical Torpedoes as a System of Defense," *Southern Historical Society Papers*, II (1876), 6.
12 *Navy Records*, VII, 61; VIII, 848–49.
13 *Harper's Weekly*, August 2, 1862; *Navy Records*, VII, 543.
14 Davidson, "Electrical Torpedoes as a System of Defense," 3.
15 R. O. Crowley, "The Confederate Torpedo Service," *Century Magazine*, XLVI (1898), 291.

Chapter III

1 Robert V. Johnson and Clarence C. Buel (eds.), *Battles and Leaders of the Civil War.* . . . (New York, 1887–88), II, 201, hereinafter cited as *Battles and Leaders.*
2 Luther S. Dickey, *History of the Eighty-fifth Regiment Pennsylvania Volunteer Infantry 1861–1865* (New York, 1915), 37.
3 These three letters, all dated May 5, 1862, appear in *History of the Fifth Massachusetts Battery* (Boston, 1902), 244–45. See also, *Harper's Weekly,* May 24, 1862.
4 Dickey, *Eighty-fifth Pennsylvania Infantry,* 37.
5 *Fifth Massachusetts Battery,* 248.
6 New York *Times,* May 7, 1862.
7 *The Bivouac,* I (1883), 52.
8 *Official Records,* XI, Pt. 3, p. 135; George B. McClellan, *McClellan's Own Story* (New York, 1887), 318.
9 Howard K. Beale (ed.), *The Diary of Edward Bates 1859–1866* (Washington, 1933), 255.
10 *Official Records,* XI, Pt. 3, p. 135; Lossing, *Pictorial History,* II, 378. See also, *Fifth Massachusetts Battery,* 244, and the New York *Times,* May 5 and 7, 1862.
11 Gabriel J. Rains, "Torpedoes," *Southern Historical Society Papers,* III (1877), 256–59; Davidson, "The Electrical Submarine Mine 1861–1865," 459.
12 J. W. Minnich, "Incidents of the Peninsular Campaign," *Confederate Veteran,* XXX (1922), 53.
13 *Official Records,* XI, Pt. 3, p. 516; Jefferson Davis, *The Rise and Fall of the Confederate Government* (New York, 1881), II, 97.
14 Minnich, "Incidents of the Peninsular Campaign," 53; *Official Records,* XI, Pt. 3, p. 509.
15 *Official Records,* XI, Pt. 3, pp. 509–11.
16 *Ibid.,* 608; Davis, *Rise and Fall of the Confederate Government,* II, 208; Rains, "Torpedoes," 260.
17 *Official Records,* XI, Pt. 3, p. 608; XVIII, 743; *Navy Records,* VII, 546.

Chapter IV

1 Victor Ernest Rudolph von Sheliha, *A Treatise on Coast-Defence: Based on the Experiences Gained by Officers of the Corps of Engineers of the Army of the Confederate States* . . . . (London, 1868), 220.
2 Danville Leadbetter to Engineer Bureau, June 4, 1863, in Vol. 7, Register of Letters Received, Confederate Engineers, Record Group 109, National Archives.
3 West Beckwith to Engineer Bureau, Oct. 8, Nov. 16, 1864, in Vol. 7, Letters Received, Confederate Engineers; C. Williams to Engineer Bureau, Dec. 28, Dec. 30, 1863, June 2, 1864, in Vol. 7, Letters Received, Confederate Engineers; *Journal of the Congress of the Confederate States of America 1861–1865* (7 vols.; Washington, 1904–1905), VII, 82, hereinafter cited as *Journal of the Confederate Congress.*
4 *Harper's Weekly,* March 15, 1862.

5   *Navy Records,* XII, 502–504; J. S. Barnes, *Submarine Warfare* (New York, 1869), 63; *Harper's Weekly,* March 15, 1862.
6   *Scientific American,* March 15, and April 17, 1862; Barnes, *Submarine Warfare,* 65.
7   *Navy Records,* X, 11.
8   Barnes, *Submarine Warfare,* 65.
9   *Official Records,* XIV, 700, 735; XIII, 757.
10 Bradford, *Torpedo Warfare,* 49; A. T. Mahan, *The Gulf and Inland Waters* (New York, 1883), 116, 118; *Navy Records,* XXIII, 545.
11 *Navy Records,* XXIII, 548.
12 *Ibid.,* 549–50. See also, Thomas O. Selfridge, *Memoirs of Thomas O. Selfridge, Jr., Rear Admiral, U.S.N.* (New York, 1924), 75.
13 Mahan, *The Gulf and Inland Waters,* 118; H. D. Brown, "The First Successful Torpedo and What It Did," *Confederate Veteran,* XVIII (1910), 169. For statements by Brown as to the type of torpedo used, see Isaac N. Brown, "Confederate Torpedoes in the Yazoo," in *Battles and Leaders,* III, 580; and Richardson (ed.), *Messages and Papers of the Confederacy,* I, 474. See also, Moore (ed.), *Rebellion Record,* VII, 253; and *Scientific American,* January 10, 1863.
14 Richardson (ed.), *Messages and Papers of the Confederacy,* I, 472–77; *Journal of the Confederate Congress,* IV, 80–81, 152–54, 222, 224, 228, 276.
15 *Navy Records,* XXIII, 547, 551. An attempt was made to salvage the wreck in 1864. The *Cairo* was located again in 1959 and her pilot house was raised in 1960. She was raised from the bottom in 1964.
16 *Ibid.,* 567–68.
17 *Ibid.,* 588, 593–605.
18 *Ibid.,* 602.
19 Brown, "Confederate Torpedoes in the Yazoo," in *Battles and Leaders,* III, 580.

Chapter v

1   The formula was revealed by Sheliha, *Treatise on Coast-Defence,* 231; by W. R. King, *Torpedoes: Their Invention and Use, From the First Application to the Art of War to the Present Time* (Washington, 1866), 14–15, and by Bradford, *Torpedo Warfare,* 52. As late as 1960, five land mines with Rains fuses were recovered near Mobile, Alabama. The powder was still quite dangerous.
2   Bradford, *Torpedo Warfare,* 55; Barnes, *Submarine Warfare,* 59; *Navy Records,* XV, 428.
3   Barnes, *Submarine Warfare,* 69. See also, Bradford, *Torpedo Warfare,* 66; King, *Torpedoes,* 10; and J. B. Jones, *A Rebel War Clerk's Diary* (Philadelphia, 1866), I, 245.
4   James Russell Soley, *The Blockade and the Cruisers* (New York, 1883), 216.
5   *Ibid.,* 216, 218; *Navy Records,* XIII, 699–731; Barnes, *Submarine Warfare,* 83–85; Bradford, *Torpedo Warfare,* 50; *Harper's Weekly,* March 28, 1863; Daniel Ammen, *The Atlantic Coast* (New York, 1883), 84–87.

6   *Harper's Weekly,* April 18, 1863; *Navy Records,* XIX, 543, 640.
7   *Navy Records,* XIX, 673, 676.
8   Jones, *Diary,* I, 245.
9   *Ibid.*
10   Rowland (ed.), *Jefferson Davis,* V, 504; *Official Records,* XVIII, 1082.
11   *Official Records,* XVIII, 1082; Rowland (ed.), *Jefferson Davis,* V, 496.
12   Rowland (ed.), *Jefferson Davis,* V, 504.
13   *Official Records,* XVIII, 1082. For the list of expenses see Gabriel Rains, Military Service Record in Carded Files, Record Group 109, National Archives.
14   Jones, *Diary,* II, 8.
15   Davis, *Rise and Fall of the Confederate Government,* II, 424–25.
16   Barnes, *Submarine Warfare,* 71–73.
17   Brown, "Confederate Torpedoes in the Yazoo," in *Battles and Leaders,* III, 580; *Navy Records,* XXV, 280.
18   *Navy Records,* XXV, 281.
19   *Ibid.,* 282–83; Barnes, *Submarine Warfare,* 90–91; Bradford, *Torpedo Warfare,* 88–91; Mahan, *The Gulf and Inland Waters,* 177; Brown, "Confederate Torpedoes in the Yazoo," in *Battles and Leaders,* III, 580.
20   *Navy Records,* XXV, 264, 282; David D. Porter, *The Naval History of the Civil War* (New York, 1886), 332.
21   *Official Records,* XXXIV, Pt. 2, p. 854.
22   Quoted in King, *Torpedoes,* 5–6.
23   *Official Records,* XXXIV, Pt. 2, p. 855.
24   Thomas O. Selfridge, "The Navy in the Red River," in *Battles and Leaders,* IV, 362.
25   *Navy Records,* XXVI, 62, 68–78.

Chapter VI

1   *Official Records,* XIV, 948–52; *Navy Records,* IX, 770; A. W. Taft, "The Signal Service Corps," *Southern Historical Society Papers,* XXV (1898), 132–33; Jones, *Diary,* I, 299; Ammen, *The Atlantic Coast,* 92–94; M. Martin Gray, Military Service Record in Carded Files, Record Group 109, National Archives.
2   *Official Records,* XVIII, 1082; *Navy Records,* XVI, 412; John Johnson; *The Defense of Charleston Harbor Including Fort Sumter and Adjacent Islands* (Charleston, 1890), clxxii.
3   *Navy Records,* XIV, 58.
4   *Official Records,* XIV, 288.
5   *Official Records,* XXVIII, Pt. 2, pp. 195, 253, 290, 291.
6   *Ibid.,* 258.
7   J. A. Hamilton, "General Stephen Elliott, Lieut. James A. Hamilton, and Elliott's Torpedo," *Southern Historical Society Papers,* X (1883), 183–84.
8   *Official Records,* XXVIII, Pt. 2, pp. 293–94; Hamilton, ". . . Elliott's Torpedo," 185; *Navy Records,* XIV, 445–48.
9   George H. Gordon, *A War Diary of Events in the War of the Great Rebellion 1863–1865* (Boston, 1882), 198–200.

10 Hamilton, ". . . Elliott's Torpedo," 184. Actually, the barrier had been constructed as a result of recent attacks by small torpedo boats.

11 Letter of October 8, 1863, in Letters Received by the Secretary of the (U.S.) Navy from South Atlantic Blockading Squadron, Record Group 45, National Archives.

12 Circular, dated September 25, 1863, in Dahlgren to Welles, October 10, 1863, in Letters Received by the Secretary of the (U.S.) Navy from South Atlantic Blockading Squadron. Dahlgren had originally planned on using divers to check these obstructions during the battle.

13 "Ambrose M'Evoy, Inventor," *Confederate Veteran*, XVI (1908), 352; *Official Records*, XXVIII, Pt. 2, p. 281.

14 *Official Records*, XXVIII, Pt. 2, pp. 213, 332.

15 Johnson, *Defense of Charleston Harbor*, 148–49; Dickey, *Eighty-fifth Pennsylvania Infantry*, 277; Gordon, *War Diary*, 212, 220; D. E. Eldridge, *The Third New Hampshire and All About It* (Boston, 1883), 368–69; New York *Times*, September 17, 1863.

16 Gordon, *War Diary*, 212.

17 Johnson, *Defense of Charleston Harbor*, 159.

18 *Official Records*, XXVIII, Pt. 2, p. 324.

19 *Ibid.*, 371–72.

20 Jones, *Diary*, II, 24, 32.

21 *Navy Records*, IX, 236, 244, 248, 373, 411.

22 *Ibid.*, 303–304, 324.

23 *Official Records*, XXVIII, Pt. 2, pp. 300, 323, 336, 380; *Navy Records*, XIV, 515, 657; West Point Museum Historical Records, United States Military Academy, West Point, New York.

24 *Official Records*, XXVIII, Pt. 2, pp. 403–404, 494.

Chapter VII

1 General Peter G. T. Beauregard, "Torpedo Service at Charleston," in *Annals of the War Written by Leading Participants North and South* (Philadelphia, 1879), 525.

2 Beauregard, "Torpedo Service at Charleston," in *Annals of the War*, 514; King, *Torpedoes*, 8.

3 Beauregard, "Torpedo Service at Charleston," in *Annals of the War*, 516–17; W. T. Glassel, "Reminiscences of Torpedo Service in Charleston Harbor," *Southern Historical Society Papers*, IV (1877), 226.

4 *Official Records*, XIV, 631; Alfred Roman, *The Military Operations of General Beauregard in the War Between the States 1861 to 1865.* (New York, 1884), II, 38.

5 *Official Records*, XIV, 636.

6 *Ibid.*

7 *Ibid.*, 648–49.

8 *Ibid.*, 670; Isaac N. Brown and William T. Glassel to Engineer Bureau, Nov. 15 and 18, 1863, in Vol. 7, Letters Received, Confederate Engineers. Prior to developing the spar torpedo, Lee designed a mechanism that was to be fixed to the sides of enemy vessels by a barbed point. It was exploded by a rope which the torpedo boat pulled as it backed off. Later experiments proved that there was no danger from explosions

caused by ramming, and he dropped the idea in favor of the simpler spar type weapon. John M. Brooke to the Secretary of the Navy, Nov. 23, 1863, in Records of the Office of Ordnance and Hydrography, C.S.N., Record Group 109, National Archives.

9   *Official Records*, XIV, 670–71.
10 *Ibid.*, 671.
11 *Ibid.*, 681, 686.
12 *Ibid.*, 694, 719, 1017–1018; Beauregard, "Torpedo Service at Charleston," in *Annals of the War*, 517.
13 *Official Records*, XIV, 761, 817–18.
14 Johnson, *Defense of Charleston Harbor*, lxiii–lxiv; *Official Records*, XIV, 885, 907, 918.
15 William H. Parker, *Recollections of a Naval Officer, 1841–1865* (New York, 1883), 306.
16 Scharf, *Confederate Navy*, 687; *Official Records*, XIV, 791.
17 Glassel, "Reminiscences of Torpedo Service in Charleston Harbor," 225–29; *Official Records*, XIV, 820.
18 Glassel, "Reminiscences of Torpedo Service in Charleston Harbor," 227.
19 *Official Records*, XIV, 821, 837.
20 Glassel, "Reminiscences of Torpedo Service in Charleston Harbor," 227–28; Scharf, *Confederate Navy*, 688–96; Johnson, *Defense of Charleston Harbor*, 75.
21 Scharf, *Confederate Navy*, 688–89.
22 *Ibid.*, 689–90; *Official Records*, XIV, 898.
23 *Official Records*, XIV, 907; Scharf, *Confederate Navy*, 690.
24 Scharf, *Confederate Navy*, 692–93; Johnson, *Defense of Charleston Harbor*, 74.
25 The petition was dated July 2, 1863. *Navy Records*, XIV, 716; Scharf, *Confederate Navy*, 695–96.
26 *Navy Records*, XIV, 274–90; Ammen, *The Atlantic Coast*, 117–20; Soley, *The Blockade and the Cruisers*, 116–20.
27 Moore (ed.), *Rebellion Record*, VII, 73.
28 Glassel, "Reminiscences of Torpedo Service in Charleston Harbor," 229.
29 *Official Records* XXVIII, Pt. 2, pp. 195, 229, 249–52, 254; Beauregard, "Torpedo Service at Charleston," in *Annals of the War*, 517.
30 F. D. Lee to Engineer Bureau, June 3, June 13, June 18, 1863, in Vol. 7, Letters Received, Confederate Engineers; *Official Records*, XIV, 965–66; XXVIII, Pt. 2, p. 254.
31 *Navy Records*, XIV, 498–99, 500; Beauregard, "Torpedo Service at Charleston," in *Annals of the War*, 517.
32 Jones, *Diary*, II, 31; *Navy Records*, XIV, 497.

Chapter VIII

1   *Navy Records*, XV, 12; Beauregard, "Torpedo Service at Charleston," in *Annals of the War*, 517. James Tomb's name is spelled variously as Tomb, Tombs, or Toomb, but he used the spelling given here. The orders quoted are from Commander W. T. Muse, C.S.S. *North Carolina*, to Glassel, September 6, 1863; Special Orders No. 186, Headquarters, Dept. of South Carolina, Georgia, and Florida, Charleston, South Caro-

lina, September 18, 1863; Tucker to Glassel, September 18 and 22, 1863. These papers were found on Glassel when he was captured and are in Letters Received by the Secretary of the (U.S.) Navy from South Atlantic Blockading Squadron.

2 Beauregard, "Torpedo Service at Charleston," in *Annals of the War*, 518; Glassel, "Reminiscences of Torpedo Service in Charleston Harbor," 231.

3 Moore (ed.), *Rebellion Record*, "Poetry and Incidents," VIII, 9–10.

4 Dahlgren to Welles, October 12, 1863, in Letters Received by the Secretary of the (U.S.) Navy from South Atlantic Blockading Squadron.

5 *Navy Records*, XV, 13; Madelene Vinton Dahlgren, *Memoir of John A. Dahlgren Rear-Admiral United States Navy* (Boston, 1882), 417.

6 Rowan to Dahlgren, October 6, 1863, and Dahlgren to Welles, October 7, 1863, in Letters Received by Secretary of the (U.S.) Navy from South Atlantic Blockading Squadron.

7 Rowan to Dahlgren, November 28, 1863, in Letters Received by Secretary of the (U.S.) Navy from South Atlantic Blockading Squadron.

8 This account is based on numerous sources. The following are the most reliable: Glassel, "Reminiscences of Torpedo Service in Charleston Harbor," 230–33; James Lachlison, "Daring Deed in Saving the David—C.S. Navy," *Confederate Veteran*, XVI (1908), 78; Johnson, *Defense of Charleston Harbor*, xxxi–xxxv; Beauregard, "Torpedo Service at Charleston," in *Annals of the War*, 518–19; *Navy Records*, XV, 10–20; Scharf, *Confederate Navy*, 760; *Official Records*, XXVIII, Pt. 2, p. 398; Ammen, *The Atlantic Coast*, 140–41; James H. Tomb, "Confederate Torpedo Boats," *Confederate Veteran*, XXXI (1923), 93; Augustin Louis Taveau to his wife, October, 1864, in Augustin Louis Taveau Papers, Duke University Library, Durham, N. C. The *New Ironsides* burned and was destroyed at Philadelphia after the war. Among her major services were the bombardments of Fort Fisher in December, 1864, and January, 1865.

9 *Navy Records*, XV, 29, 30, 48, 64.

10 *Ibid.*, XIX, 631; XX, 696, 697, 705, 848–49; *Official Records*, XXXV, Pt. 1, p. 548.

11 *Official Records*, XXVIII, Pt. 2, pp. 442–43.

12 *Ibid.*, 504.

13 *Ibid.*, 605–606; 595–97.

14 *Ibid.*, 566–67, 576.

15 *Ibid*, XXXV, Pt. 1, pp. 537–38, 548.

Chapter IX

1 Issue of June 7, 1862.

2 Herbert C. Fyfe, *Submarine Warfare Past and Present* (London, 1907), 188.

3 Goldsborough to Welles, October 20, 1861, in Letters Received by Secretary of the (U.S.) Navy from North Atlantic Blockading Squadron, Record Group 45, National Archives; *Navy Records*, VI, 346–50.

4 Allan Pinkerton, *The Spy of the Rebellion; Being a True History of the Spy System of the United States Army During the Late Rebellion* (Chi-

cago, 1883), 395–403; *Navy Records*, VI, 363, 393, 304a; IX, 411; *Harper's Weekly*, November 2, 1861.

5 *Navy Records*, IX, 411; XXII, 104.

6 There is some confusion as to the submarine's exact size. See McClintock to Matthew Fontaine Maury, undated, in Vol. 46, Maury Papers, Manuscript Division, Library of Congress; Gordon Levey, "Torpedo Boats at Louisiana Soldier's Home," *Confederate Veteran*, XVII (1909), 459; William Robinson, Jr., *The Confederate Privateers* (New Haven, 1928), 166–74; *Navy Records*, Ser. II, Vol. I, 399, 401.

7 McClintock to Maury, undated, Maury Papers.

8 Scharf, *Confederate Navy*, 750.

9 McClintock to Maury, undated, Maury Papers. In 1879, a submarine was found near New Orleans by a channel dredge whose crew pulled it ashore. It lay in the mud for another generation. On April 10, 1909, the Beauregard Camp of the United Sons of Confederate Veterans had it placed on the Bank of Bayou Têche at the Louisiana State Home for Confederate Veterans. Later, the submarine was moved to Jackson Square, New Orleans, and is now outside the Louisiana State Museum.

10 McClintock to Maury, undated, Maury Papers; W. A. Alexander, "Thrilling Chapter in the History of the Confederate States Navy. Work of Submarine Boats," *Southern Historical Society Papers*, XXX (1902), 165, hereinafter cited as Alexander, "Work of Submarine Boats"; Levey, "Torpedo Boats at Louisiana Soldier's Home," 459; Robinson, *The Confederate Privateers*, 176–77.

11 *Navy Records*, XXVI, 188; XV, 229; Alexander, "Work of Submarine Boats," 165–66; Beauregard, "Torpedo Service at Charleston," in *Annals of the War*, 519; McClintock to Maury, undated, Maury Papers.

12 *Navy Records*, XV, 229; Robinson, *The Confederate Privateers*, 176; Beauregard, "Torpedo Service at Charleston," in *Annals of the War*, 519.

13 Alexander, "Work of Submarine Boats," 167.

14 E. Willis, "Torpedoes and Torpedo Boats and Their Bold and Hazardous Expeditions," (MS in Manuscript Division, Library of Congress); *Official Records*, XXVIII, Pt. 2, pp. 265, 285; Beauregard, "Torpedo Service at Charleston," in *Annals of the War*, 519; Levey, "Torpedo Boats at Louisiana Soldier's Home," 459.

15 Beauregard, "Torpedo Service at Charleston," in *Annals of the War*, 519; Alexander, "Work of Submarine Boats," 168; Levey, "Torpedo Boats at Louisiana Soldier's Home," 459; Arthur P. Ford, "The First Submarine Boat," *Confederate Veteran*, XIV (1906), 563.

16 Alexander, "Work of Submarine Boats," 169; *Navy Records*, XV, 692; Ford, "The First Submarine Boat," 564; Levey, "Torpedo Boats at Louisiana Soldier's Home," 459.

17 Beauregard, "Torpedo Service at Charleston," in *Annals of the War*, 519; *Navy Records*, XV, 692; Alexander, "Work of Submarine Boats," 168–70.

18 Alexander, "Work of Submarine Boats," 169–73; *Official Records*, XXVIII, Pt. 2, p. 553.

19 Alexander, "Work of Submarine Boats," 168–74; *Navy Records*, XV, 328–36; McClintock to Maury, undated, Maury Papers; Beauregard,

"Torpedo Service at Charleston," in *Annals of the War,* 520; Johnson, *Defense of Charleston Harbor,* cxlix.

Chapter x

1   Jones, *Diary,* I, 6.
2   *Navy Records,* IX, 145–49; Hunter Davidson to Jefferson Davis, Dec. 5, 1881, *Southern Historical Society Papers,* XXIV (1896), 286; Barnes, *Submarine Warfare,* 92; *Harper's Weekly,* Aug. 29, 1863. R. O. Crowley claimed the operator "lost his presence of mind" and fired the torpedo too soon. "Confederate Torpedo Service," *Century Magazine,* XLVI (1898), 292.
3   *Navy Records,* IX, 725.
4   Barnes, *Submarine Warfare,* 96.
5   *Ibid.,* 96–100.
6   Rowland (ed.), *Jefferson Davis,* VII, 108.
7   Scharf, *Confederate Navy,* 732, 763; *Navy Records,* X, 9–16; Crowley, "Confederate Torpedo Service," 293–94.
8   *Navy Records,* X, 51–53, 92, 267, 635–45; *Official Records,* XXXVI, Pt. 2, p. 49.
9   Rowland (ed.), *Jefferson Davis,* VII, 110.
10 William Burns, Jr., to Stephen V. Benet, Jacksonville, Florida, March 30, 1864, in West Point Museum Historical Records, West Point, New York.
11 *Official Records,* XXXV, Pt. 1, pp. 115, 369–70, 381, 584; Pt. 2, p. 397; *Navy Records,* XVI, 420–21.
12 *Official Records,* XXXV, Pt. 1, pp. 115, 388; XV, 314; Norfolk (Virginia) *New Regime,* April 25, 1864.
13 *Official Records,* XXXV, Pt. 1, p. 117; *Navy Records,* XVI, 425; Gordon, *War Diary,* 295.
14 Norfolk (Virginia) *New Regime,* April 25, 1864.
15 *Navy Records,* XV, 426; *Official Records,* XXXV, Pt. 2, pp. 116, 121, 171, 447.

Chapter xi

1   *Official Records,* XXXV, Pt. 1, p. 517.
2   Quoted in John S. Elmore to John M. Otley, Jan. 5, 1864, in Vol. 7½, Letters Received, Confederate Engineers; *Official Records,* XXXV, Pt. 1, pp. 517, 538–39.
3   Dahlgren, *Memoir,* 438; Gordon, *War Diary,* 268–69; George H. Gordon to A. H. Bowman, Feb. 3, 1864, in West Point Museum Historical Records.
4   *Official Records,* XXXV, Pt. 1, pp. 594, 597, 616–17. Gray's Journal is preserved in Letters Received by the Secretary of the (U.S.) Navy from South Atlantic Blockading Squadron.
5   These paragraphs are based on requisitions for materials submitted by Captain Gray and General Rains, found in their Military Service Records in Carded Files, Record Group 109, National Archives.
6   *Ibid.*
7   *Official Records,* XXXV, Pt. 1, pp. 155–56, Pt. 2, pp. 439, 499, 502, 503; *Navy Records,* XVI, 418–27.

8 Dahlgren, *Memoir*, 465; *Navy Records*, XV, 437–39, 638; XVI, 14; *Official Records*, XXXV, Pt. 1, pp. 14–15, 17, 238, 240; Pt. 2, p. 186.

9 *Official Records*, XXXV, Pt. 1, pp. 546, 548, 597, 598, 603; Jones, *Diary*, II, 138.

10 *Navy Records*, XV, 356–58; *Official Records*, XXXV, Pt. 2, pp. 345–46; Lachlison, "Daring Deed in Saving the David," 78; Tomb, "Confederate Torpedo Boats," 93.

11 *Navy Records*, IX, 561; XV, 397; *Official Records*, XXXV, Pt. 2, p. 359.

12 *Navy Records*, XV, 408; *Official Records*, XXXV, Pt. 2, pp. 396, 402, 406, 408.

13 *Navy Records*, IX, 510.

14 John M. Batten, *Reminiscences of Two Years in the United States Navy* (Lancaster, Pa., 1881) 14.

15 Crowley, "Confederate Torpedo Service," 297.

16 Davidson, "The Electrical Submarine Mine 1861–1865," 458; Batten, *Two Years in the United States Navy*, 133; *Navy Records*, IX, 603.

17 *Scientific American*, April 30, 1864.

18 *Journal of the Confederate Congress*, IV, 202; Rowland (ed.), *Jefferson Davis*, VII, 387.

19 *Navy Records*, X, 616, 621; Norfolk (Virginia) *New Regime*, April 23, 1864.

20 *Official Records*, XXXV, Pt. 2, pp. 460, 504; *Navy Records*, XV, 408, 678.

21 *Navy Records*, X, 112–13.

Chapter XII

1 Morris Schaff, "The Explosion at City Point," in *Civil War Papers Read Before the Commandery of the State of Massachusetts, Military Order of the Loyal Legion of the United States* (Boston, 1900), II, 479.

2 Horace Porter, *Campaigning with Grant* (New York, 1879), 273.

3 Schaff, "The Explosion at City Point," in *Civil War Papers*, 482; New York *Times*, August 12, 1864.

4 Schaff, "The Explosion at City Point," in *Civil War Papers*, 479–80.

5 Horace Porter, *Campaigning with Grant*, 273; Joseph P. Farley, *West Point in the Early Sixties with Incidents of the War* (Troy, N.Y., 1902), 147.

6 Schaff, "The Explosion at City Point," in *Civil War Papers*, 481; New York *Times*, August 12, 1864.

7 Schaff, "The Explosion at City Point," in *Civil War Papers*, 480–82; Farley, *West Point in the Sixties*, 149.

8 Schaff, "The Explosion at City Point," in *Civil War Papers*, 487; Farley, *West Point in the Sixties*, 147; Horace Porter, *Campaigning with Grant*, 274.

9 Schaff, "The Explosion at City Point," in *Civil War Papers*, 483.

10 After the war, Maxwell's report was found in Richmond by the Federals and was sent to Secretary of War Edwin M. Stanton by Major General Henry W. Halleck on June 3, 1865. It is found in Schaff, "The Explosion at City Point," in *Civil War Papers*, 483–84 and in Farley, *West Point in the Sixties*, 148–49. In 1873, when Colonel

Orville E. Babcock was President Grant's secretary, he claimed that Maxwell appeared before him to complain about ill treatment regarding a patent case and, at that time, told the secretary of the explosive used at City Point. Horace Porter, *Campaigning with Grant*, 274.

11 *Navy Records*, XXVI, 190; James A. Shutt and Joseph Thatcher, "The Courtenay Coal Torpedo," *Military Collector and Historian*, XI (1959), 7–8.

12 Courtenay to Jefferson Davis, December 7, 1863, in Shutt and Thatcher, "The Courtenay Coal Torpedo," 7–8.

13 *Navy Records*, XXVI, 186–87, 313; Courtenay to Col. H. E. Clark, January 19, 1864, in Shutt and Thatcher, "The Courtenay Coal Torpedo," 7–8; Robinson, *The Confederate Privateers*, 326.

14 Shutt and Thatcher, "The Courtenay Coal Torpedo," 7–8; *Navy Records*, XXVI, 184.

15 *Harper's Weekly*, April 30, 1864; *Scientific American*, April 30, 1864, July 28, 1864; Bradford, *Torpedo Warfare*, 62.

16 New York *Times*, May 18, 1865; *Navy Records*, XII, 136.

17 Bradford, *Torpedo Warfare*, 62; New York *Times*, May 18, 1865; Shutt and Thatcher, "The Courtenay Coal Torpedo," 7–8.

18 David D. Porter, *Incidents and Anecdotes of the Civil War* (New York, 1886) 264–66.

19 *Harper's Weekly*, December 17, 1864.

Chapter XIII

1 *Navy Records*, X, 771, 791. These wooden torpedo boats were 6 feet, 3 inches wide and 3 feet, 9 inches deep. *Ibid.*, XI, 707.

2 *Navy Records*, XI, 748–56; *Harper's Weekly*, February 11, 1865. It is probable that a small torpedo in the collection of the Richmond National Battlefield Park is one of these models.

3 *Navy Records*, XI, 757–77, 206–207; Brooke to Rains, December 8, 1864, in Records of the Office of Ordnance and Hydrography, C.S.N., Record Group 109, National Archives.

4 *Navy Records*, XI, 120.

5 *Ibid.*, 124, 199.

6 Scharf, *Confederate Navy*, 740.

7 *Navy Records*, XI, 632–41.

8 David D. Porter to W. A. Parker, Jan. 26, 1865, in Letters Received by Secretary of the (U.S.) Navy from North Atlantic Blockading Squadron.

9 *Navy Records*, XI, 634–46; 662–63.

10 *Ibid.*, 706–707.

11 *Navy Records*, XII, 10–12.

12 *Ibid.*, XI, 717.

13 Scharf, *Confederate Navy*, 742.

14 *Ibid.*, 743–44.

Chapter XIV

1 Batten, *Two Years in the United States Navy*, 65, 67; *Navy Records*, XI, 160–65, 176; Barnes, *Submarine Warfare*, 108; New York *Herald*, January 9, 1865.

2 *Navy Records,* XI, 163.
3 *Ibid.,* 164; New York *Herald,* January 9, 1865; J. C. Ives to (C.S.) Secretary of War, October 1, 1863, in Vol. 7, Letters Received, Confederate Engineers.
4 *Navy Records,* XI, 165.
5 *Ibid.,* 168–69.
6 *Ibid.,* 179–82.
7 *Ibid.,* 725; XII, 8, 9, 17–18.
8 *Navy Records,* XII, 107, 108, 116, 150, 160.
9 *Ibid.,* V, 412, 414.
10 Foxhall A. Parker to Gideon Welles, May 11, 1864, in Letters Received by Secretary of the (U.S.) Navy from the Potomac Flotilla, Record Group 45, National Archives.
11 *Navy Records,* V, 421–25.
12 *Ibid.,* 432.
13 Alonzo G. Draper to Foxhall Parker, June 22, 1864, in Letters Received by Secretary of the (U.S.) Navy from the Potomac Flotilla.
14 *Ibid.; Navy Records,* V, 436–39.
15 Parker to Welles, November 22, 1864, in Letters Received by Secretary of the (U.S.) Navy from the Potomac Flotilla.
16 Ralph J. Roske and Charles Van Doren in *Lincoln's Commando; The Biography of William B. Cushing, U.S.N.* (New York, 1957), give a full treatment of this incident. An example of the thin line between heroism and disgrace are the cases of the young ensigns who took these boats down the bay. Ensign W. L. Howorth completed the trip and was with Cushing when the *Albemarle* was sunk. He was given honorable mention for the feat. The unfortunate Stockholm was dismissed from the Navy on November 7, 1864.
17 *Navy Records,* V, 497–501, XI, 382.

Chapter xv

1 This narrative is based on a number of accounts: Bradford, *Torpedo Warfare,* 102–109; Loyall Farragut, *The Life of David Glasgow Farragut, First Admiral of the United States Navy* . . . (New York, 1879), 411–20; *Official Records,* XXXIX, Pt. 2, pp. 707–708, 739, 756, 759, 786; Gordon, *War Diary,* 342; *Navy Records,* XXI, 404, 405, 417, 438, 556, 569, 598; *Battles and Leaders,* IV, 388, 391; Scharf, *Confederate Navy,* 556, 561; *Southern Historical Society Papers,* IX (1881), 471; *Harper's Weekly,* September 10, 1864; Mahan, *The Gulf and Inland Waters,* 218–50.
   There has been much discussion as to just what Farragut said at Mobile. Some authorities do not believe he uttered the famous phrase, "Damn the torpedoes!"; others think he did. For discussion, see Clarence E. MacCarthney, *Mr. Lincoln's Admirals* (New York, 1956); Charles Lee Lewis, *David Glasgow Farragut, Our First Admiral* (Annapolis, 1943); and Virgil C. Jones, *The Final Effort,* Volume III of *The Civil War at Sea* (New York, 1962). At the time of the sinking of the *Tecumseh,* Farragut used the phrase, "go ahead" in signals from the *Hartford. Navy Records,* XXI, 508.

2 Bradford, *Torpedo Warfare*, 57–58. Barrett was the first to notify Rains of the sinking of the *Tecumseh.*

3 *Official Records*, XXXIX, Pt. 2, pp. 768–82; Barnes, *Submarine Warfare*, 106; *Harper's Weekly*, September 24, 1864.

4 W. G. Jones to J. S. Palmer, December 10, 1864, in Letters Received by Secretary of the (U.S.) Navy from West Gulf Blockading Squadron, Record Group 45, National Archives.

Chapter XVI

1 Willis, "Torpedoes and Torpedo Boats," (MS in Manuscript Division, Library of Congress).

2 *Official Records*, XXXV, Pt. 2, p. 154; XXXVIII, Pt. 4, p. 579; Rains's order is in Special Orders No. 133, June 8, 1864, Gabriel Rains, Military Service Record in Carded Files.

3 *Official Records*, XLIV, 865–66, 880, 885, 934; Rowland (ed.), *Jefferson Davis*, V, 407, 410, VIII, 416.

4 Gordon, *War Diary*, 268.

5 *Official Records*, XLIV, 61, 79, 110; *Navy Records*, XVI, 362; James G. Crozier, 26th Iowa Infantry to "Lue," December 14, 1864, in West Point Museum Historical Records; Charles C. Jones, Jr., *Historical Sketch of the Chatham Artillery During the Confederate Struggle for Independence* (Albany, N.Y., 1867), 140–43.

6 *Navy Records*, XVI, 164, 218.

7 *Ibid.*, 374–75, 380–89, 418. Gray was born in Delaware in 1821 and had lived in North Carolina. There is nothing in his file at the National Archives about his imprisonment or court martial. However, there are several requisitions for rope: for 647 pounds Sept. 12, 1863, and 421 pounds October 20. The latter document bears the notation that it was "absolutely necessary to fill my orders." M. Martin Gray, Military Service Record in Carded Files.

8 *Official Records*, XXXV, Pt. 2, p. 648.

9 *Navy Records*, XVI, 171–79; *Official Records*, XLVII, Pt. 1, pp. 1068, 1134–35.

Chapter XVII

1 *Navy Records*, XVI, 339, 342, 372, 378, 387; *Harper's New Monthly Magazine*, December, 1870, p. 12.

2 *Navy Records*, XVI, 374–78, 379, 380, 411–18; Johnson, *Defense of Charleston Harbor*, clxxi.

3 Dahlgren, *Memoir*, 504; *Navy Records*, XVI, 283–84.

4 Barnes, *Submarine Warfare*, 113–14.

5 *Navy Records*, XVI, 296, 408–409.

6 King, *Torpedoes*, 1–3.

7 Barnes, *Submarine Warfare*, 112, 168–71.

8 *Navy Records*, XII, 44; Porter, *Incidents and Anecdotes of the Civil War*, 278; Bradford, *Torpedo Warfare*, 113; Scharf, *Confederate Navy*, 767.

9 Porter, *Incidents and Anecdotes of the Civil War*, 267–77.

Chapter XVIII

1 *Navy Records*, XII, 41, 184, 186.
2 *Ibid.*, 185–86.
3 *Navy Records*, Series II, Vol. II, 688–89, 724–25, 770.
4 Bradford, *Torpedo Warfare*, 59; King, *Torpedoes*, 16; Barnes, *Submarine Warfare*, 74. Several examples of Confederate torpedoes were collected by Captain Peter S. Michie, U.S.A., and sent to the West Point Museum.
5 *Official Records*, XLII, Pt. 3, pp. 215–16, 282, 1219.
6 George Alfred Townsend, *Rustics in Rebellion; a Yankee on the Road to Richmond, 1861–1865* (Chapel Hill, 1950), 265–66.
7 *Navy Records*, XII, 97–104.
8 "Torpedoes, Bureau of Ordnance, Navy," Vol. 1, pp. 236–37 in Records of the Bureau of Ordnance, Record Group 74, National Archives.
9 Porter, *Incidents and Anecdotes of the Civil War*, 309.
10 Rowland (ed.), *Jefferson Davis*, IX, 230–31.
11 *Navy Records*, V, 575, 576.
12 Sheliha's Military Service Record in Carded Files, Record Group 109, National Archives, contains a number of letters from the Prussian Consul at New Orleans, General Buckner, and others.
13 *Navy Records*, XXII, 267–69; D. H. Maury to Samuel Cooper, Feb. 3, 1865, *Southern Historical Society Papers*, IX (1881), 81.
14 Wirt Armstrong Cate (ed.) *Two Soldiers: the Campaign Diaries of Thomas J. Key, C.S.A.; December 7, 1863–May 17, 1865, and Robert J. Campbell, U.S.A.; January 1, 1864–July 21, 1864* (Chapel Hill, 1938), 188; Arthur H. Burnham, to Superintendent, U.S. Military Academy, October 21, 1865, in West Point Museum Historical Records.
15 *Harper's Weekly*, February 25, 1865; D. H. Maury to Samuel Cooper, February 3, 1865, *Southern Historical Society Papers*, IX (1881), 81; *Navy Records*, XX, 269.
16 V. Sheliha to Engineer Bureau, Aug. 16, Aug. 18, Aug. 20, 1864, in Vols. 7, 7½, Letters Received, Confederate Engineers; Sheliha, Military Service Record in Carded Files.
17 *Navy Records*, XXII, 133.
18 "Report of Submarine Operations, February, 1865, at Mobile," and "Property Return of Submarine Defenses of Mobile, Jan.–Feb., 1865," in Records of the Confederate Engineers, Record Group 109, National Archives. The *Althea* was eventually raised and repaired.
19 *Navy Records*, XXII, 71.
20 *Ibid.*
21 *Ibid.*, 72; Barnes, *Submarine Warfare*, 116–17. Like the *Milwaukee* the *Osage* was built in St. Louis by Eads. She was commissioned July 10, 1863.
22 *Navy Records*, XXII, 72–73; Barnes, *Submarine Warfare*, 118.
23 Chicago *Tribune*, April 10, 1865.
24 King, *Torpedoes*, 4–6.
25 H. U. Dowd Diary (MS in the Jackson County Historical Society, Independence, Mo.). Not all the mines were found. Three were recovered as late as 1960.

26 *Navy Records*, XXII, 86–88; Barnes, *Submarine Warfare*, 120.
27 Frank E. Alward, "A Sailor's Service," *The Maine Bugle*, II (1895), 212; *Navy Records*, XXII, 131; Barnes, *Submarine Warfare*, 119.
28 Barnes, *Submarine Warfare*, 119; *Navy Records*, XXII, 130.
29 *Navy Records*, XXII, 131.
30 Scharf, *Confederate Navy*, 767.
31 *Navy Records*, XXVI, 516, 524, 525, 569, 706–13, 751.
32 *Ibid.*, XXVII, 47, 48, 86–87.
33 *Ibid.*, 228, 230.

Chapter XIX

1 Letters Patent No. 3154, December 8, 1865, to Nathaniel John Holmes, British Patent Office; U.S. Minister to Mexico to Secretary of State, September 10, 1865, October 28, 1865, in Diplomatic Correspondence, Record Group 59, National Archives.
2 Rowland (ed.), *Jefferson Davis*, VII, 107–109, 387–91; Davidson, "Electrical Torpedoes as a System of Defense," 2–6; Davidson to Jefferson Davis, Dec. 5, 1881, *Southern Historical Society Papers*, XXIV (1896), 285–86. Davidson, "The Electrical Submarine Mine 1861–1865," 456–58.
3 St. Louis City Directories for 1874 and 1906.
4 Scharf, *Confederate Navy*, 754; John W. Dubose, "Lieutenant William T. Glassel, of Alabama," *Confederate Veteran*, XXIV (1916), 193; W. T. Glassel, "Torpedo Service in Charleston Harbor," *Confederate Veteran*, XXV (1917), 113–15.
5 Scharf, *Confederate Navy*, 306.
6 Charleston *News and Courier*, August 29, 1885.
7 *Ibid.;* see also the *Missouri Republican*, August 19, 1876, and February 25, 1877; Merchant's Exchange Collection and miscellaneous St. Louis papers in Missouri Historical Society (St. Louis).
8 Shutt and Thatcher, "The Courtenay Coal Torpedo," 7–8.
9 *Battles and Leaders*, II, 89; Charles R. Kennon to Milton F. Perry, July 22, 1961.
10 Lossing, *Pictorial History*, III, 562; King, *Torpedoes*, 90.

The first sources one should consult for material relating to Confederate torpedoes are the studies made by officers of United States services. These are: J. S. Barnes, *Submarine Warfare* (New York, 1869); W. R. King, *Torpedoes: Their Invention and Use, From the First Application to the Art of War to the Present Time* (Washington, 1866), and Lieutenant Commander Royal B. Bradford, U.S.N., *History of Torpedo Warfare* (Newport, R.I., 1882). These are technical studies which include few discussions of personalities or dramatizations other than those in quoted accounts. They provide sound studies of the workings of the many models of torpedoes and their capabilities.

I was fortunate to have used these sources in conjunction with the only remaining important collection of Confederate torpedoes: that at the West Point Museum, United States Military Academy, West Point, New York. Not only is an examination of the workmanship, materials, and features of these century-old weapons instructive, but many of the drawings in the above-mentioned books were based on these items and on the notes and sketches of Peter S. Michie, who collected many of them. In addition, the stories behind the discoveries of the torpedoes, which are found in the museum's files, are used in this narrative.

Notes on Sources

To supplement the technical sources I made liberal use of letters and reports found in those Civil War treasure troves, the *Official Records of the Union and Confederate Navies in the War of the Rebellion* (31 vols.; Washington, 1894–1927) and *The War of the Rebellion: A Compilation of the Official Records of the Union and Confederate Armies* (70 vols. in 128 parts, Washington, 1880–1901).

Personal narratives are scattered through the *Southern Historical Society Papers* (47 vols., Richmond, 1876–1930), the *Confederate Veteran*, and Robert V. Johnson and Clarence C. Buel (eds.), *Battles and Leaders of the Civil War* (4 vols., New York, 1887–88). Those used are cited in the notes. Other diaries and book-length accounts of value are: J. B. Jones, *A Rebel War Clerk's Diary* (2 vols., Philadelphia, 1866), which is full of the rumors and gossip of wartime Richmond; Admiral David D. Porter, *The Naval History of the Civil War of the Great Rebellion 1863–1865* (New York, 1886), a sprightly and interesting book; and Captain William H. Parker, *Recollections of a Naval Officer 1841–1865* (New York, 1883)—probably the most entertaining book of them all.

Among the unpublished sources of special value was Betty Herndon Maury's diary in the Manuscript Division of the Library of Congress, which contains firsthand accounts from Matthew Fontaine Maury; and the letters sent to the Secretary of the U.S. Navy by the commanders of various squadrons and flotillas, found in Record Group 45 in the National Archives. These letters include a number of unpublished notes and sometimes are at variance with the published versions in the *Navy Records*.

Official unpublished papers from the Confederate side are scarce, and one must "dig" for every reference. The records of the torpedo bureaus were destroyed—some intentionally. A fortunate find was the book of letters sent to the Navy's Bureau of Ordnance and Hydrography for November, 1864–March, 1865, now in Records of the Office of Ordnance and Hydrography, C.S.N., Record Group 109, National Archives. Confederate Engineer Records are few, but one can reconstruct many of them from the three Registers of Letters Received in the same record group.

A most interesting item is James McClintock's letter to Matthew F. Maury telling of his experiences with the *Hunley*. It is one of the few important unpublished accounts concerning this ship and can be found in Volume 46 of the Maury Papers, Manuscript Division, Library of Congress. Another interesting unpublished paper is one by E. Willis, "Torpedoes and Torpedo Boats and Their Bold and Hazardous Expeditions" also in the Manuscript Division, Library of Congress. It was evidently written for publication, and its author apparently witnessed many of the events described.

Most of all, I value memories: the pleasant conversation with Mrs. N. M. Osborne, Maury's granddaughter; riding about the waters of Hampton Roads and the James River, tracing the routes of Maury's attacks; savoring the warm peacefulness of the sun-speckled waters of the tree-lined Roanoke; standing upon the crumbling trenches of Fort Anderson amidst the stillness of the forest; poring over charts and maps; breathing the salt-laden air on the glistening beaches outside Mobile Bay; and disassembling mines and torpedoes that had been made by Rains, Lee, and Singer. Such sources cannot be fully described in any bibliography.